Text Mining

Text Mining
Applications and Theory

Michael W. Berry

University of Tennessee, USA

Jacob Kogan

University of Maryland Baltimore County, USA

A John Wiley and Sons, Ltd., Publication

Library of Congress Cataloguing-in-Publication Data

Berry, Michael W.
    Text mining : applications and theory / Michael Berry, Jacob Kogan.
        p.   cm.
    Includes bibliographical references and index.
    ISBN 978-0-470-74982-1 (cloth)
    1. Data mining – Congresses. 2. Natural language processing (Computer science) – Congresses.
I. Kogan, Jacob, 1954- II. Title.
    QA76.9.D343B467 2010
    006.3'12 – dc22

                                                                                2010000137

A catalogue record for this book is available from the British Library.

ISBN: 978-0-470-74982-1

Typeset in 10/12 Times-Roman by Laserwords Private Limited, Chennai, India
Printed and bound in Great Britain by TJ International Ltd, Padstow, Cornwall.

Contents

List of Contributors   xi

Preface   xiii

PART I   TEXT EXTRACTION, CLASSIFICATION, AND CLUSTERING   **1**

1   Automatic keyword extraction from individual documents   **3**
   1.1   Introduction   3
     1.1.1   Keyword extraction methods   4
   1.2   Rapid automatic keyword extraction   5
     1.2.1   Candidate keywords   6
     1.2.2   Keyword scores   7
     1.2.3   Adjoining keywords   8
     1.2.4   Extracted keywords   8
   1.3   Benchmark evaluation   9
     1.3.1   Evaluating precision and recall   9
     1.3.2   Evaluating efficiency   10
   1.4   Stoplist generation   11
   1.5   Evaluation on news articles   15
     1.5.1   The MPQA Corpus   15
     1.5.2   Extracting keywords from news articles   15
   1.6   Summary   18
   1.7   Acknowledgements   19
   References   19

2   Algebraic techniques for multilingual document clustering   **21**
   2.1   Introduction   21
   2.2   Background   22
   2.3   Experimental setup   23
   2.4   Multilingual LSA   25
   2.5   Tucker1 method   27

2.6   PARAFAC2 method   28
2.7   LSA with term alignments   29
2.8   Latent morpho-semantic analysis (LMSA)   32
2.9   LMSA with term alignments   33
2.10   Discussion of results and techniques   33
2.11   Acknowledgements   35
References   35

**3   Content-based spam email classification using
      machine-learning algorithms**   **37**
3.1   Introduction   37
3.2   Machine-learning algorithms   39
       3.2.1   Naive Bayes   39
       3.2.2   LogitBoost   40
       3.2.3   Support vector machines   41
       3.2.4   Augmented latent semantic indexing spaces   43
       3.2.5   Radial basis function networks   44
3.3   Data preprocessing   45
       3.3.1   Feature selection   45
       3.3.2   Message representation   47
3.4   Evaluation of email classification   48
3.5   Experiments   49
       3.5.1   Experiments with PU1   49
       3.5.2   Experiments with ZH1   51
3.6   Characteristics of classifiers   53
3.7   Concluding remarks   54
3.8   Acknowledgements   55
References   55

**4   Utilizing nonnegative matrix factorization for email
      classification problems**   **57**
4.1   Introduction   57
       4.1.1   Related work   59
       4.1.2   Synopsis   60
4.2   Background   60
       4.2.1   Nonnegative matrix factorization   60
       4.2.2   Algorithms for computing NMF   61
       4.2.3   Datasets   63
       4.2.4   Interpretation   64
4.3   NMF initialization based on feature ranking   65
       4.3.1   Feature subset selection   66
       4.3.2   FS initialization   66
4.4   NMF-based classification methods   70
       4.4.1   Classification using basis features   70
       4.4.2   Generalizing LSI based on NMF   72

4.5   Conclusions   78
4.6   Acknowledgements   79
References   79

5   Constrained clustering with k-means type algorithms   81
5.1   Introduction   81
5.2   Notations and classical k-means   82
5.3   Constrained k-means with Bregman divergences   84
      5.3.1   Quadratic k-means with cannot-link constraints   84
      5.3.2   Elimination of must-link constraints   87
      5.3.3   Clustering with Bregman divergences   89
5.4   Constrained smoka type clustering   92
5.5   Constrained spherical k-means   95
      5.5.1   Spherical k-means with cannot-link constraints only   96
      5.5.2   Spherical k-means with cannot-link and must-link
              constraints   98
5.6   Numerical experiments   99
      5.6.1   Quadratic k-means   100
      5.6.2   Spherical k-means   100
5.7   Conclusion   101
References   102

PART II   ANOMALY AND TREND DETECTION   105

6   Survey of text visualization techniques   107
6.1   Visualization in text analysis   107
6.2   Tag clouds   108
6.3   Authorship and change tracking   110
6.4   Data exploration and the search for novel patterns   111
6.5   Sentiment tracking   111
6.6   Visual analytics and FutureLens   113
6.7   Scenario discovery   114
      6.7.1   Scenarios   115
      6.7.2   Evaluating solutions   115
6.8   Earlier prototype   116
6.9   Features of FutureLens   117
6.10 Scenario discovery example: bioterrorism   119
6.11 Scenario discovery example: drug trafficking   121
6.12 Future work   123
References   126

7   Adaptive threshold setting for novelty mining   129
7.1   Introduction   129
7.2   Adaptive threshold setting in novelty mining   131

7.2.1   Background   131
7.2.2   Motivation   132
7.2.3   Gaussian-based adaptive threshold setting   132
7.2.4   Implementation issues   137
7.3   Experimental study   138
7.3.1   Datasets   138
7.3.2   Working example   139
7.3.3   Experiments and results   142
7.4   Conclusion   146
References   147

8   **Text mining and cybercrime**   **149**
8.1   Introduction   149
8.2   Current research in Internet predation and cyberbullying   151
8.2.1   Capturing IM and IRC chat   151
8.2.2   Current collections for use in analysis   152
8.2.3   Analysis of IM and IRC chat   153
8.2.4   Internet predation detection   153
8.2.5   Cyberbullying detection   158
8.2.6   Legal issues   159
8.3   Commercial software for monitoring chat   159
8.4   Conclusions and future directions   161
8.5   Acknowledgements   162
References   162

PART III   TEXT STREAMS   **165**

9   **Events and trends in text streams**   **167**
9.1   Introduction   167
9.2   Text streams   169
9.3   Feature extraction and data reduction   170
9.4   Event detection   171
9.5   Trend detection   174
9.6   Event and trend descriptions   176
9.7   Discussion   180
9.8   Summary   181
9.9   Acknowledgements   181
References   181

10 **Embedding semantics in LDA topic models**   **183**
10.1 Introduction   183
10.2 Background   184

    10.2.1 Vector space modeling   184
    10.2.2 Latent semantic analysis   185
    10.2.3 Probabilistic latent semantic analysis   185
10.3 Latent Dirichlet allocation   186
    10.3.1 Graphical model and generative process   187
    10.3.2 Posterior inference   187
    10.3.3 Online latent Dirichlet allocation (OLDA)   189
    10.3.4 Illustrative example   191
10.4 Embedding external semantics from Wikipedia   193
    10.4.1 Related Wikipedia articles   194
    10.4.2 Wikipedia-influenced topic model   194
10.5 Data-driven semantic embedding   194
    10.5.1 Generative process with data-driven semantic embedding   195
    10.5.2 OLDA algorithm with data-driven semantic embedding   196
    10.5.3 Experimental design   197
    10.5.4 Experimental results   199
10.6 Related work   202
10.7 Conclusion and future work   202
References   203

Index   **205**

List of Contributors

Loulwah AlSumait
Department of Information Science
Kuwait University, Kuwait.
lalsumai@gmu.edu

Brett W. Bader
Sandia National Laboratories
Albuquerque, NM, USA.
Bwbader@sandia.gov

Daniel Barbará
Department of Computer Science
George Mason University
Fairfax, VA, USA
dbarbara@gmu.edu

Michael W. Berry
University of Tennessee
Min H. Kao Department of Electrical
    Engineering and Computer Science
Knoxville, TN, USA.
berry@eecs.utk.edu

Peter A. Chew
Sandia National Laboratories
Albuquerque, NM, USA.
pchew@sandia.gov

Wendy Cowley
Pacific Northwest National Laboratory
Richland, WA, USA.
wendy@pnl.gov

Nick Cramer
Pacific Northwest National Laboratory
Richland, WA, USA.
nick.cramer@pnl.gov

Carlotta Domeniconi
Department of Computer Science
George Mason University
Fairfax, VA, USA.
carlotta@cs.gmu.edu

Lynne Edwards
Department of Media and
    Communication Studies
Ursinus College
Collegeville, PA, USA.
edwards@ursinus.edu

Dave Engel
Pacific Northwest National Laboratory
Richland, WA, USA.
dave.engel@pnl.gov

Wilfried N. Gansterer
Research Lab Computational
    Technologies and Applications
University of Vienna, Austria.
wilfried.gansterer@univie.ac.at

Andreas G. K. Janecek
Research Lab Computational
    Technologies and Applications
University of Vienna, Austria.
andreas.janecek@univie.ac.at

Eric P. Jiang
University of San Diego
San Diego, CA, USA.
jiang@sandiego.edu

Jacob Kogan
University of Maryland, Baltimore
    County
Baltimore, MD, USA.
kogan@umbc.edu

April Kontostathis
Department of Mathematics and
    Computer Science
Ursinus College
Collegeville, PA, USA.
akontostathis@ursinus.edu

Amanda Leatherman
Department of Media and
    Communication Studies
Ursinus College
Collegeville, PA, USA.
aleat001@umaryland.edu

Charles Nicholas
University of Maryland, Baltimore
    County
Baltimore, MD, USA.
nicholas@umbc.edu

Andrey A. Puretskiy
University of Tennessee
Min H. Kao Department of Electrical
    Engineering and Computer Science
Knoxville, TN, USA.
puretski@eecs.utk.edu

Stuart Rose
Pacific Northwest National Laboratory
Richland, WA, USA.
stuart.rose@pnl.gov

Gregory L. Shutt
University of Tennessee
Min H. Kao Department of Electrical
    Engineering and Computer Science
Knoxville, TN, USA.
Shutt@eecs.utk.edu

Ziqiu Su
University of Maryland, Baltimore
    County
Baltimore, MD, USA.
ziqiu1@umbc.edu

Wenyin Tang
Nanyang Technological University
Singapore.
wenyintang@ntu.edu.sg

Flora S. Tsai
Nanyang Technological University
Singapore
efstsai@ntu.edu.sg

Pu Wang
Department of Computer Science
George Mason University
Fairfax, VA
pwang7@gmu.edu

Paul Whitney
Pacific Northwest National Laboratory
Richland, WA
paul.whitney@pnl.gov

Preface

The proliferation of digital computing devices and their use in communication continues to result in an increased demand for systems and algorithms capable of mining textual data. Thus, the development of techniques for mining unstructured, semi-structured, and fully structured textual data has become quite important in both academia and industry. As a result, a one-day workshop on text mining was held on May 2, 2009 in conjunction with the SIAM Ninth International Conference on Data Mining to bring together researchers from a variety of disciplines to present their current approaches and results in text mining. The workshop surveyed the emerging field of text mining, the application of techniques of machine learning in conjunction with natural language processing, information extraction, and algebraic/mathematical approaches to computational information retrieval. Many issues are being addressed in this field ranging from the development of new document classification and clustering models to novel approaches for topic detection, tracking, and visualization.

With over 40 applied mathematicians and computer scientists representing universities, industrial corporations, and government laboratories from six different countries, the workshop featured both invited and contributed talks on the use of techniques from machine learning, knowledge discovery, natural language processing, and information retrieval to design computational models for automated text analysis and mining. Most of the invited and contributed papers presented at the workshop have been compiled and expanded for this volume. Collectively, they span several major topic areas in text mining:

1. Keyword extraction

2. Classification and clustering

3. Anomaly and trend detection

4. Text streams.

This volume presents state-of-the-art algorithms for text mining from both the academic and industrial perspectives. Each chapter is self-contained and is completed by a list of references. A subject-level index is also provided at the end of the volume. Familiarity with basic undergraduate-level mathematics is needed for several of the chapters. The volume should be useful for a novice to the field as well as for an expert in text mining research.

The inherent differences in the words written by authors and those used by readers continue to fuel the development of effective search and retrieval algorithms and software in the field of text mining. This volume demonstrates how advancements in the fields of applied mathematics, computer science, machine learning, and natural language processing can collectively capture, classify, and interpret words and their contexts. The words alone are not enough.

Michael W. Berry and Jacob Kogan
Knoxville, TN and Baltimore, MD
August 2009
www.wiley.com/go/berry_mining

Part I

TEXT EXTRACTION, CLASSIFICATION, AND CLUSTERING

1

Automatic keyword extraction from individual documents

Stuart Rose, Dave Engel, Nick Cramer and Wendy Cowley

1.1 Introduction

Keywords, which we define as a sequence of one or more words, provide a compact representation of a document's content. Ideally, keywords represent in condensed form the essential content of a document. Keywords are widely used to define queries within information retrieval (IR) systems as they are easy to define, revise, remember, and share. In comparison to mathematical signatures, keywords are independent of any corpus and can be applied across multiple corpora and IR systems.

Keywords have also been applied to improve the functionality of IR systems. Jones and Paynter (2002) describe Phrasier, a system that lists documents related to a primary document's keywords, and that supports the use of keyword anchors as hyperlinks between documents, enabling a user to quickly access related material. Gutwin et al. (1999) describe Keyphind, which uses keywords from documents as the basic building block for an IR system. Keywords can also be used to enrich the presentation of search results. Hulth (2004) describes Keegle, a system that dynamically provides keyword extracts for web pages returned from a Google search. Andrade and Valencia (1998) present a system that automatically annotates protein function with keywords extracted from the scientific literature that are associated with a given protein.

Text Mining: Applications and Theory   edited by Michael W. Berry and Jacob Kogan
© 2010, John Wiley & Sons, Ltd

1.1.1   Keyword extraction methods

Despite their utility for analysis, indexing, and retrieval, most documents do not have assigned keywords. Most existing approaches focus on the manual assignment of keywords by professional curators who may use a fixed taxonomy, or rely on the authors' judgment to provide a representative list. Research has therefore focused on methods to automatically extract keywords from documents as an aid either to suggest keywords for a professional indexer or to generate summary features for documents that would otherwise be inaccessible.

Early approaches to automatically extract keywords focus on evaluating corpus-oriented statistics of individual words. Jones (1972) and Salton et al. (1975) describe positive results of selecting for an index vocabulary the statistically discriminating words across a corpus. Later keyword extraction research applies these metrics to select discriminating words as keywords for individual documents. For example, Andrade and Valencia (1998) base their approach on comparison of word frequency distributions within a text against distributions from a reference corpus.

While some keywords are likely to be evaluated as statistically discriminating within the corpus, keywords that occur in many documents within the corpus are not likely to be selected as statistically discriminating. Corpus-oriented methods also typically operate only on single words. This further limits the measurement of statistically discriminating words because single words are often used in multiple and different contexts.

To avoid these drawbacks, we focus our interest on methods of keyword extraction that operate on individual documents. Such document-oriented methods will extract the same keywords from a document regardless of the current state of a corpus. Document-oriented methods therefore provide context-independent document features, enabling additional analytic methods such as those described in Engel et al. (2009) and Whitney et al. (2009) that characterize changes within a text stream over time. These document-oriented methods are suited to corpora that change, such as collections of published technical abstracts that grow over time or streams of news articles. Furthermore, by operating on a single document, these methods inherently scale to vast collections and can be applied in many contexts to enrich IR systems and analysis tools.

Previous work on document-oriented methods of keyword extraction has combined natural language processing approaches to identify part-of-speech (POS) tags that are combined with supervised learning, machine-learning algorithms, or statistical methods.

Hulth (2003) compares the effectiveness of three term selection approaches: noun-phrase (NP) chunks, n-grams, and POS tags, with four discriminative features of these terms as inputs for automatic keyword extraction using a supervised machine-learning algorithm.

Mihalcea and Tarau (2004) describe a system that applies a series of syntactic filters to identify POS tags that are used to select words to evaluate as keywords. Co-occurrences of the selected words within a fixed-size sliding window are accumulated within a word co-occurrence graph. A graph-based ranking algorithm (TextRank) is applied to rank words based on their associations in the graph, and then top ranking words are selected as keywords. Keywords that are adjacent in the document are combined to form multi-word keywords. Mihalcea and Tarau (2004) report that TextRank achieves its best performance when only nouns and adjectives are selected as potential keywords.

Matsuo and Ishizuka (2004) apply a chi-square measure to calculate how selectively words and phrases co-occur within the same sentences as a particular subset of frequent terms in the document text. The chi-square measure is applied to determine the bias of word co-occurrences in the document text which is then used to rank words and phrases as keywords of the document. Matsuo and Ishizuka (2004) state that the degree of biases is not reliable when term frequency is small. The authors present an evaluation on full text articles and a working example on a 27-page document, showing that their method operates effectively on large documents.

In the following sections, we describe Rapid Automatic Keyword Extraction (RAKE), an unsupervised, domain-independent, and language-independent method for extracting keywords from individual documents. We provide details of the algorithm and its configuration parameters, and present results on a benchmark dataset of technical abstracts, showing that RAKE is more computationally efficient than TextRank while achieving higher precision and comparable recall scores. We then describe a novel method for generating stoplists, which we use to configure RAKE for specific domains and corpora. Finally, we apply RAKE to a corpus of news articles and define metrics for evaluating the exclusivity, essentiality, and generality of extracted keywords, enabling a system to identify keywords that are essential or general to documents in the absence of manual annotations.

1.2   Rapid automatic keyword extraction

In developing RAKE, our motivation has been to develop a keyword extraction method that is extremely efficient, operates on individual documents to enable application to dynamic collections, is easily applied to new domains, and operates well on multiple types of documents, particularly those that do not follow specific grammar conventions. Figure 1.1 contains the title and text for a typical abstract, as well as its manually assigned keywords.

RAKE is based on our observation that keywords frequently contain multiple words but rarely contain standard punctuation or stop words, such as the function words *and*, *the*, and *of*, or other words with minimal lexical meaning. Reviewing the manually assigned keywords for the abstract in Figure 1.1, there is only one keyword that contains a stop word (*of* in *set of natural numbers*). Stop words are typically dropped from indexes within IR systems and not included in various text analyses as they are considered to be uninformative or meaningless. This reasoning is based on the expectation that such words are too frequently and broadly used to aid users in their analyses or search tasks. Words that do

Compatibility of systems of linear constraints over the set of natural numbers

Criteria of compatibility of a system of linear Diophantine equations, strict inequations, and nonstrict inequations are considered. Upper bounds for components of a minimal set of solutions and algorithms of construction of minimal generating sets of solutions for all types of systems are given. These criteria and the corresponding algorithms for constructing a minimal supporting set of solutions can be used in solving all the considered types of systems and systems of mixed types.

Manually assigned keywords:
linear constraints, set of natural numbers, linear Diophantine equations, strict inequations, nonstrict inequations, upper bounds, minimal generating sets

Figure 1.1   A sample abstract from the Inspec test set and its manually assigned keywords.

carry meaning within a document are described as content bearing and are often referred to as content words.

The input parameters for RAKE comprise a list of stop words (or stoplist), a set of phrase delimiters, and a set of word delimiters. RAKE uses stop words and phrase delimiters to partition the document text into candidate keywords, which are sequences of content words as they occur in the text. Co-occurrences of words within these candidate keywords are meaningful and allow us to identify word co-occurrence without the application of an arbitrarily sized sliding window. Word associations are thus measured in a manner that automatically adapts to the style and content of the text, enabling adaptive and fine-grained measurement of word co-occurrences that will be used to score candidate keywords.

1.2.1   Candidate keywords

RAKE begins keyword extraction on a document by parsing its text into a set of candidate keywords. First, the document text is split into an array of words by the specified word delimiters. This array is then split into sequences of contiguous words at phrase delimiters and stop word positions. Words within a sequence are assigned the same position in the text and together are considered a candidate keyword.

Figure 1.2 shows the candidate keywords in the order that they are parsed from the sample technical abstract shown in Figure 1.1. The candidate keyword

Compatibility – systems – linear constraints – set – natural numbers – Criteria – compatibility – system – linear Diophantine equations – strict inequations – nonstrict inequations – Upper bounds – components – minimal set – solutions – algorithms – minimal generating sets – solutions – systems – criteria – corresponding algorithms – constructing – minimal supporting set – solving – systems – systems

Figure 1.2   Candidate keywords parsed from the sample abstract.

linear Diophantine equations begins after the stop word *of* and ends with a comma. The following word *strict* begins the next candidate keyword *strict inequations*.

1.2.2   Keyword scores

After every candidate keyword is identified and the graph of word co-occurrences (shown in Figure 1.3) is complete, a score is calculated for each candidate keyword and defined as the sum of its member word scores. We evaluated several metrics for calculating word scores, based on the degree and frequency of word vertices in the graph: (1) word frequency ($freq(w)$), (2) word degree ($deg(w)$), and (3) ratio of degree to frequency ($deg(w)/freq(w)$).

The metric scores for each of the content words in the sample abstract are listed in Figure 1.4. In summary, $deg(w)$ favors words that occur often and in longer candidate keywords; $deg(minimal)$ scores higher than $deg(systems)$. Words that occur frequently regardless of the number of words with which they co-occur are favored by $freq(w)$; $freq(systems)$ scores higher than $freq(minimal)$. Words that predominantly occur in longer candidate keywords are favored by $deg(w)/freq(w)$; $deg(diophantine)/freq(diophantine)$ scores higher than $deg(linear)/freq(linear)$. The score for each candidate keyword is computed as the sum of its member

	algorithms	bounds	compatibility	components	constraints	constructing	corresponding	criteria	diophantine	equations	generating	inequations	linear	minimal	natural	nonstrict	numbers	set	sets	solving	strict	supporting	system	systems	upper
algorithms	2						1																		
bounds		1																							1
compatibility			2																						
components				1																					
constraints					1								1												
constructing						1																			
corresponding	1						1																		
criteria								2																	
diophantine									1	1			1												
equations									1	1			1												
generating											1			1					1						
inequations												2				1					1				
linear					1				1	1			2												
minimal											1			3				2	1			1			
natural															1		1								
nonstrict												1				1									
numbers															1		1								
set														2				3				1			
sets											1			1					1						
solving																				1					
strict												1									1				
supporting														1				1				1			
system																							1		
systems																								4	
upper		1																							1

Figure 1.3   The word co-occurrence graph for content words in the sample abstract.

	algorithms	bounds	compatibility	components	constraints	constructing	corresponding	criteria	diophantine	equations	generating	inequations	linear	minimal	natural	nonstrict	numbers	set	sets	solving	strict	supporting	system	systems	upper
deg(w)	3	2	2	1	2	1	2	2	3	3	3	4	5	8	2	2	2	6	3	1	2	3	1	4	2
freq(w)	2	1	2	1	1	1	1	2	1	1	1	2	2	3	1	1	1	3	1	1	1	1	1	4	1
deg(w) / freq(w)	1.5	2	1	1	2	1	2	1	3	3	3	2	2.5	2.7	2	2	2	2	3	1	2	3	1	1	2

Figure 1.4   Word scores calculated from the word co-occurrence graph.

minimal generating sets (8.7), linear diophantine equations (8.5), minimal supporting set (7.7), minimal set (4.7), linear constraints (4.5), natural numbers (4), strict inequations (4), nonstrict inequations (4), upper bounds (4), corresponding algorithms (3.5), set (2), algorithms (1.5), compatibility (1), systems (1), criteria (1), system (1), components (1),constructing (1), solving (1)

Figure 1.5   Candidate keywords and their calculated scores.

word scores. Figure 1.5 lists each candidate keyword from the sample abstract using the metric *deg(w)/freq(w)* to calculate individual word scores.

1.2.3   Adjoining keywords

Because RAKE splits candidate keywords by stop words, extracted keywords do not contain interior stop words. While RAKE has generated strong interest due to its ability to pick out highly specific terminology, an interest was also expressed in identifying keywords that contain interior stop words such as *axis of evil*. To find these RAKE looks for pairs of keywords that adjoin one another at least twice in the same document and in the same order. A new candidate keyword is then created as a combination of those keywords and their interior stop words. The score for the new keyword is the sum of its member keyword scores.

It should be noted that relatively few of these linked keywords are extracted, which adds to their significance. Because adjoining keywords must occur twice in the same order within the document, their extraction is more common on texts that are longer than short abstracts.

1.2.4   Extracted keywords

After candidate keywords are scored, the top T scoring candidates are selected as keywords for the document. We compute T as one-third the number of words in the graph, as in Mihalcea and Tarau (2004).

The sample abstract contains 28 content words, resulting in $T = 9$ keywords. Table 1.1 lists the keywords extracted by RAKE compared to the sample abstract's manually assigned keywords. We use the statistical measures precision, recall and F-measure to evaluate the accuracy of RAKE. Out of nine keywords extracted, six are true positives; that is, they exactly match six of the manually assigned keywords. Although *natural numbers* is similar to the assigned

Table 1.1   Comparison of keywords extracted by RAKE to manually assigned keywords for the sample abstract.

Extracted by RAKE	Manually assigned
minimal generating sets	minimal generating sets
linear diophantine equations	linear Diophantine equations
minimal supporting set	
minimal set	
linear constraints	linear constraints
natural numbers	
strict inequations	strict inequations
nonstrict inequations	nonstrict inequations
upper bounds	upper bounds
	set of natural numbers

keyword *set of natural numbers*, for the purposes of the benchmark evaluation it is considered a miss. There are therefore three false positives in the set of extracted keywords, resulting in a precision of 67%. Comparing the six true positives within the set of extracted keywords to the total of seven manually assigned keywords results in a recall of 86%. Equally weighting precision and recall generates an F-measure of 75%.

1.3   Benchmark evaluation

To evaluate performance we tested RAKE against a collection of technical abstracts used in the keyword extraction experiments reported in Hulth (2003) and Mihalcea and Tarau (2004), mainly for the purpose of allowing direct comparison with their results.

1.3.1   Evaluating precision and recall

The collection consists of 2000 Inspec abstracts for journal papers from Computer Science and Information Technology. The abstracts are divided into a training set with 1000 abstracts, a validation set with 500 abstracts, and a testing set with 500 abstracts. We followed the approach described in Mihalcea and Tarau (2004), using the testing set for evaluation because RAKE does not require a training set. Extracted keywords for each abstract are compared against the abstract's associated set of manually assigned uncontrolled keywords.

Table 1.2 details RAKE's performance using a generated stoplist, Fox's stoplist (Fox 1989), and T as one-third the number of words in the graph. For each method, which corresponds to a row in the table, the following information is shown: the total number of extracted keywords and mean per abstract; the number of correct extracted keywords and mean per abstract; precision; recall; and F-measure. Results published within Hulth (2003) and Mihalcea and Tarau

Table 1.2   Results of automatic keyword extraction on 500 abstracts in the
Inspec test set using RAKE, TextRank (Mihalcea and Tarau 2004) and
supervised learning (Hulth 2003).

Method	Extracted keywords		Correct keywords		Precision	Recall	F-measure
	Total	Mean	Total	Mean			
RAKE ($T = 0.33$)							
KA stoplist ($df > 10$)	6052	12.1	2037	4.1	**33.7**	41.5	**37.2**
Fox stoplist	7893	15.8	2054	4.2	26	42.2	32.1
TextRank							
Undirected, co-occ. window = 2	6784	13.6	2116	4.2	31.2	43.1	36.2
Undirected, co-occ. window = 3	6715	13.4	1897	3.8	28.2	38.6	32.6
(Hulth 2003)							
Ngram with tag	7815	15.6	1973	3.9	25.2	**51.7**	33.9
NP chunks with tag	4788	9.6	1421	2.8	29.7	37.2	33
Pattern with tag	7012	14	1523	3	21.7	39.9	28.1

the, and, of, a, in, is, for, to, we, this, are, with, as, on, it, an, that, which, by, using, can,
paper, from, be, based, has, was, have, or, at, such, also, but, results, proposed, show,
new, these, used, however, our, were, when, one, not, two, study, present, its, sub, both,
then, been, they, all, presented, if, each, approach, where, may, some, more, use,
between, into, 1, under, while, over, many, through, addition, well, first, will, there,
propose, than, their, 2, most, sup, developed, particular, provides, including, other, how,
without, during, article, application, only, called, what, since, order, experimental, any

Figure 1.6   Top 100 words in the generated stoplist.

(2004) are included for comparison. The highest values for precision, recall, and
F-measure are shown in bold. As noted, perfect precision is not possible with
any of the techniques as the manually assigned keywords do not always appear
in the abstract text. The highest precision and F-measure are achieved using
RAKE with a generated stoplist based on keyword adjacency, a subset of which
is listed in Figure 1.6. With this stoplist RAKE yields the best results in terms of
F-measure and precision, and provides comparable recall. With Fox's stoplist,
RAKE achieves a high recall while experiencing a drop in precision.

1.3.2   Evaluating efficiency

Because of increasing interest in energy conservation in large data centers, we
also evaluated the computational cost associated with extracting keywords with
RAKE and TextRank. TextRank applies syntactic filters to a document text to

identify content words and accumulates a graph of word co-occurrences in a window size of 2. A rank for each word in the graph is calculated through a series of iterations until convergence below a threshold is achieved.

We set TextRank's damping factor $d = 0.85$ and its convergence threshold to 0.0001, as recommended in Mihalcea and Tarau (2004). We do not have access to the syntactic filters referenced in Mihalcea and Tarau (2004), so were unable to evaluate their computational cost.

To minimize disparity, all parsing stages in the respective extraction methods are identical, TextRank accumulates co-occurrences in a window of size 2, and RAKE accumulates word co-occurrences within candidate keywords. After co-occurrences are tallied, the algorithms compute keyword scores according to their respective methods. The benchmark was implemented in Java and executed in the Java SE Runtime Environment (JRE) 6 on a Dell Precision T7400 workstation.

We calculated the total time for RAKE and TextRank (as an average over 100 iterations) to extract keywords from the Inspec testing set of 500 abstracts, after the abstracts were read from files and loaded in memory. RAKE extracted keywords from the 500 abstracts in 160 milliseconds. TextRank extracted keywords in 1002 milliseconds, over 6 times the time of RAKE.

Referring to Figure 1.7, we can see that as the number of content words for a document increases, the performance advantage of RAKE over TextRank increases. This is due to RAKE's ability to score keywords in a single pass whereas TextRank requires repeated iterations to achieve convergence on word ranks.

Based on this benchmark evaluation, it is clear that RAKE effectively extracts keywords and outperforms the current state of the art in terms of precision, efficiency, and simplicity. As RAKE can be put to use in many different systems and applications, in the next section we discuss a method for stoplist generation that may be used to configure RAKE on particular corpora, domains, and languages.

1.4   Stoplist generation

Stoplists are widely used in IR and text analysis applications. However, there is remarkably little information describing methods for their creation. Fox (1989) presents an analysis of stoplists, noting discrepancies between stated conventions and actual instances and implementations of stoplists. The lack of technical rigor associated with the creation of stoplists presents a challenge when comparing text analysis methods. In practice, stoplists are often based on common function words and hand-tuned for particular applications, domains, or specific languages.

We evaluated the use of term frequency as a metric for automatically selecting words for a stoplist. Table 1.3 lists the top 50 words by term frequency in the training set of abstracts in the benchmark dataset. Additional metrics shown for each word are document frequency, adjacency frequency, and keyword frequency. Adjacency frequency reflects the number of times the word occurred adjacent to

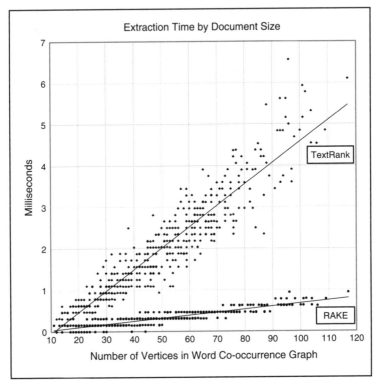

Figure 1.7   Comparison of TextRank and RAKE extraction times on individual documents.

an abstract's keywords. Keyword frequency reflects the number of times the word occurred within an abstract's keywords.

Looking at the top 50 frequent words, in addition to the typical function words, we can see that *system*, *control*, and *method* are highly frequent within technical abstracts and highly frequent within the abstracts' keywords. Selecting solely by term frequency will therefore cause content-bearing words to be added to the stoplist, particularly if the corpus of documents is focused on a particular domain or topic. In those circumstances, selecting stop words by term frequency presents a risk of removing important content-bearing words from analysis.

We therefore present the following method for automatically generating a stoplist from a set of documents for which keywords are defined. The algorithm is based on the intuition that words adjacent to, and not within, keywords are less likely to be meaningful and therefore are good choices for stop words.

To generate our stoplist we identified for each abstract in the Inspec training set the words occurring adjacent to words in the abstract's uncontrolled keyword list. The frequency of each word occurring adjacent to a keyword was accumulated across the abstracts. Words that occurred more frequently within keywords than adjacent to them were excluded from the stoplist.

Table 1.3 The 50 most frequent words in the Inspec training set listed in descending order by term frequency.

Word	Term frequency	Document frequency	Adjacency frequency	Keyword frequency
the	8611	978	3492	3
of	5546	939	1546	68
and	3644	911	2104	23
a	3599	893	1451	2
to	3000	879	792	10
in	2656	837	1402	7
is	1974	757	1175	0
for	1912	767	951	9
that	1129	590	330	0
with	1065	577	535	3
are	1049	576	555	1
this	964	581	645	0
on	919	550	340	8
an	856	501	332	0
we	822	388	731	0
by	773	475	283	0
as	743	435	344	0
be	595	395	170	0
it	560	369	339	13
system	**507**	**255**	**86**	**202**
can	452	319	250	0
based	451	293	168	15
from	447	309	187	0
using	428	282	260	0
control	**409**	**166**	**12**	**237**
which	402	280	285	0
paper	398	339	196	1
systems	**384**	**194**	**44**	**191**
method	**347**	**188**	**78**	**85**
data	**347**	**159**	**39**	**131**
time	**345**	**201**	**24**	**95**
model	**343**	**157**	**37**	**122**
information	**322**	**153**	**18**	**151**
or	315	218	146	0
s	314	196	27	0
have	301	219	149	0
has	297	225	166	0
at	296	216	141	0
new	294	197	93	4
two	287	205	83	5

(continued overleaf)

Table 1.3   (*Continued*)

Word	Term frequency	Document frequency	Adjacency frequency	Keyword frequency
algorithm	**267**	**123**	**36**	**96**
results	262	221	129	14
used	262	204	92	0
was	254	125	161	0
these	252	200	93	0
also	251	219	139	0
such	249	198	140	0
problem	**234**	**137**	**36**	**55**
design	**225**	**110**	**38**	**68**

To evaluate this method of generating stoplists, we created six stoplists, three of which select words for the stoplist by term frequency (TF), and three which select words by term frequency but also exclude words from the stoplist whose keyword frequency was greater than their keyword adjacency frequency. We refer to this latter set of stoplists as keyword adjacency (KA) stoplists since they primarily include words that are adjacent to and not within keywords.

Table 1.4   Comparison of RAKE performance using stoplists based on term frequency (TF) and keyword adjacency (KA).

Method	Stoplist size	Extracted keywords		Correct keywords		Precision	Recall	*F*-measure
		Total	Mean	Total	Mean			
RAKE ($T = 0.33$) TF stoplist ($df > 10$)	1347	3670	7.3	606	1.2	16.5	12.3	14.1
TF stoplist ($df > 25$)	527	5563	11.1	1032	2.1	18.6	21.0	19.7
TF stoplist ($df > 50$)	205	7249	14.5	1520	3.0	21.0	30.9	25.0
RAKE ($T = 0.33$) KA stoplist ($df > 10$)	763	6052	12.1	2037	4.1	33.7	41.5	37.2
KA stoplist ($df > 25$)	325	7079	14.2	2103	4.3	29.7	42.8	35.1
KA stoplist ($df > 50$)	147	8013	16.0	2117	4.3	26.4	43.1	32.8

Each of the stoplists was set as the input stoplist for RAKE, which was then run on the testing set of the Inspec corpus of technical abstracts. Table 1.4 lists the precision, recall, and F-measure for the keywords extracted by each of these runs. The KA stoplists generated by our method outperformed the TF stoplists generated by term frequency. A notable difference between results achieved using the two types of stoplists is evident in Table 1.4: the F-measure improves as more words are added to a KA stoplist, whereas when more words are added to a TF stoplist the F-measure degrades. Furthermore, the best TF stoplist underperforms the worst KA stoplist. This verifies that our algorithm for generating stoplists is adding the right stop words and excluding content words from the stoplist.

Because the generated KA stoplists leverage manually assigned keywords, we envision that an ideal application would be within existing digital libraries or IR systems and collections where defined keywords exist or are easily identified for a subset of the documents. Stoplists only need to be generated once for particular domains, enabling RAKE to be applied to new and future articles, facilitating the annotation and indexing of new documents.

1.5   Evaluation on news articles

While we have shown that a simple set of configuration parameters enables RAKE to efficiently extract keywords from individual documents, it is worth investigating how well extracted keywords represent the essential content within a corpus of documents for which keywords have not been manually assigned. The following section presents results on application of RAKE to the Multi-Perspective Question Answering (MPQA) Corpus (CERATOPS 2009).

1.5.1   The MPQA Corpus

The MPQA Corpus consists of 535 news articles provided by the Center for the Extraction and Summarization of Events and Opinions in Text (CERATOPS). Articles in the MPQA Corpus are from 187 different foreign and US news sources and date from June 2001 to May 2002.

1.5.2   Extracting keywords from news articles

We extracted keywords from title and text fields of documents in the MPQA Corpus and set a minimum document threshold of two because we are interested in keywords that are associated with multiple documents.

Candidate keyword scores were based on word scores as $deg(w)/freq(w)$ and as $deg(w)$. Calculating word scores as $deg(w)/freq(w)$, RAKE extracted 517 keywords referenced by an average of 4.9 documents. Calculating word scores as $deg(w)$, RAKE extracted 711 keywords referenced by an average of 8.1 documents.

This difference in average number of referenced document counts is the result of longer keywords having lower frequency across documents. The metric *deg(w)/freq(w)* favors longer keywords and therefore results in extracted keywords that occur in fewer documents in the MPQA Corpus.

In many cases a subject is occasionally presented in its long form and more frequently referenced in its shorter form. For example, referring to Table 1.5, *kyoto protocol on climate change* and *1997 kyoto protocol* occur less frequently than the shorter *kyoto protocol*. Because our interest in the analysis of news articles is to connect articles that reference related content, we set RAKE to score words by *deg(w)* in order to favor shorter keywords that occur across more documents.

Because most documents are unique within any given corpus, we expect to find variability in what documents are essentially about as well as how each document represents specific subjects. While some documents may be primarily about the *kyoto protocol*, *greenhouse gas emissions*, and *climate change*, other documents may only make references to those subjects. Documents in the former set will likely have *kyoto protocol*, *greenhouse gas emissions*, and *climate change* extracted as keywords whereas documents in the latter set will not.

In many applications, users have a desire to capture all references to extracted keywords. For the purposes of evaluating extracted keywords, we accumulate

Table 1.5   Keywords extracted with word scores by *deg(w)* and *deg(w)/freq(w)*.

Keyword	Scored by *deg(w)*		Scored by *deg(w)/freq(w)*	
	edf(w)	*rdf(w)*	*edf(w)*	*rdf(w)*
kyoto protocol legally obliged developed countries	2	2	2	2
eu leader urge russia to ratify kyoto protocol	2	2	2	2
kyoto protocol on climate change	2	2	2	2
ratify kyoto protocol	2	2	2	2
kyoto protocol requires	2	2	2	2
1997 kyoto protocol	2	4	4	4
kyoto protocol	31	44	7	44
kyoto	10	12	–	–
kyoto accord	3	3	–	–
kyoto pact	2	3	–	–
sign kyoto protocol	2	2	–	–
ratification of the kyoto protocol	2	2	–	–
ratify the kyoto protocol	2	2	–	–
kyoto agreement	2	2	–	–

counts on how often each extracted keyword is referenced by documents in the corpus. The referenced document frequency of a keyword, *rdf(k)*, is the number of documents in which the keyword occurred as a candidate keyword. The extracted document frequency of a keyword, *edf(k)*, is the number of documents from which the keyword was extracted.

A keyword that is extracted from all of the documents in which it is referenced can be characterized as *exclusive* or *essential*, whereas a keyword that is referenced in many documents but extracted from a few may be characterized as *general*. Comparing the relationship of *edf(k)* and *rdf(k)* allows us to characterize the exclusivity of a particular keyword. We therefore define keyword exclusivity *exc(k)* as shown in Equation (1.1):

$$exc(k) = \frac{edf(k)}{rdf(k)}. \tag{1.1}$$

Of the 711 extracted keywords, 395 have an exclusivity score of 1, indicating that they were extracted from every document in which they were referenced. Within that set of 395 exclusive keywords, some occur in more documents than others and can therefore be considered more essential to the corpus of documents. In order to measure how essential a keyword is, we define the essentiality of a keyword, *ess(k)*, as shown in Equation (1.2):

$$ess(k) = exc(k) \times edf(k). \tag{1.2}$$

Figure 1.8 lists the top 50 essential keywords extracted from the MPQA corpus, listed in descending order by their *ess(k)* scores. According to CERATOPS, the MPQA corpus comprises 10 primary topics, listed in Table 1.6, which are well represented by the 50 most essential keywords as extracted and ranked by RAKE.

In addition to keywords that are essential to documents, we can also characterize keywords by how general they are to the corpus. In other words, how

---

united states (32), human rights (24), kyoto protocol (22), international space station (18), mugabe (16), space station (14), human rights report (12), greenhouse gas emissions (12), chavez (11), taiwan issue (11), president chavez (10), human rights violations (10), president bush (10), palestinian people (10), prisoners of war (9), president hugo chavez (9), kyoto (8), taiwan (8), israeli government (8), hugo chavez (8), climate change (8), space (8), axis of evil (7), president fernando henrique cardoso (7), palestinian (7), palestinian territories (6), taiwan strait (6), russian news agency interfax (6), prisoners (6), taiwan relations act (6), president robert mugabe (6), presidential election (6), geneva convention (5), palestinian authority (5), venezuelan president hugo chavez (5), chinese president jiang zemin (5), opposition leader morgan tsvangirai (5), french news agency afp (5), bush (5), north korea (5), camp x-ray (5), rights (5), election (5), mainland china (5), al qaeda (5), president (4), south africa (4), global warming (4), bush administration (4), mdc leader (4)

Figure 1.8   Top 50 essential keywords from the MPQA Corpus, with corresponding ess(k) *score in parentheses.*

Table 1.6   MPQA Corpus topics and definitions.

Topic	Description
argentina	Economic collapse in Argentina
axisofevil	Reaction to President Bush's 2002 State of the Union Address
guantanamo	US holding prisoners in Guantanamo Bay
humanrights	Reaction to US State Department report on human rights
kyoto	Ratification of Kyoto Protocol
mugabe	2002 Presidential election in Zimbabwe
settlements	Israeli settlements in Gaza and West Bank
spacestation	Space missions of various countries
taiwan	Relations between Taiwan and China
venezuela	Presidential coup in Venezuela

government (147), countries (141), people (125), world (105), report (91), war (85), united states (79), china (71), president (69), iran (60), bush (56), japan (50), law (44), peace (44), policy (43), officials (43), israel (41), zimbabwe (39), taliban (36), prisoners (35), opposition (35), plan (35), president george (34), axis (34), administration (33), detainees (32), treatment (32), states (30), european union (30), palestinians (30), election (29), rights (28), international community (27), military (27), argentina (27), america (27), guantanamo bay (26), official (26), weapons (24), source (24), eu (23), attacks (23), united nations (22), middle east (22), bush administration (22), human rights (21), base (20), minister (20), party (19), north korea (18)

Figure 1.9   Top 50 general keywords from the MPQA Corpus, with corresponding gen(k) score in parentheses.

often was a keyword referenced by documents from which it was not extracted? In this case we define generality of a keyword, *gen(k)*, as shown in Equation (1.3):

$$gen(k) = rdf(k) \times (1.0 - exc(k)). \tag{1.3}$$

Figure 1.9 lists the top 50 general keywords extracted from the MPQA corpus, listed in descending order by their *gen(k)* scores. It should be noted that general keywords and essential keywords are not mutually exclusive. Within the top 50 for both metrics, there are several shared keywords: *united states*, *president*, *bush*, *prisoners*, *election*, *rights*, *bush administration*, *human rights*, and *north korea*. Keywords that are both highly essential and highly general are essential to a set of documents within the corpus but also referenced by a significantly greater number of documents within the corpus than other keywords.

1.6   Summary

We have shown that our automatic keyword extraction technology, RAKE, achieves higher precision and similar recall in comparison to existing techniques.

In contrast to methods that depend on natural language processing techniques to achieve their results, RAKE takes a simple set of input parameters and automatically extracts keywords in a single pass, making it suitable for a wide range of documents and collections.

Finally, RAKE's simplicity and efficiency enable its use in many applications where keywords can be leveraged. Based on the variety and volume of existing collections and the rate at which documents are created and collected, RAKE provides advantages and frees computing resources for other analytic methods.

1.7   Acknowledgements

This work was supported by the National Visualization and Analytics Center™ (NVAC™), which is sponsored by the US Department of Homeland Security Program and located at the Pacific Northwest National Laboratory (PNNL), and by Laboratory Directed Research and Development at PNNL. PNNL is managed for the US Department of Energy by Battelle Memorial Institute under Contract DE-AC05-76RL01830.

We also thank Anette Hulth, for making available the dataset used in her experiments.

References

Andrade M and Valencia A 1998 Automatic extraction of keywords from scientific text: application to the knowledge domain of protein families. *Bioinformatics* **14**(7), 600–607.

CERATOPS 2009 MPQA Corpus http://www.cs.pitt.edu/mpqa/ceratops/corpora.html.

Engel D, Whitney P, Calapristi A and Brockman F 2009 Mining for emerging technologies within text streams and documents. *Proceedings of the Ninth SIAM International Conference on Data Mining*. Society for Industrial and Applied Mathematics.

Fox C 1989 A stop list for general text. *ACM SIGIR Forum*, vol. 24, pp. 19–21. ACM, New York, USA.

Gutwin C, Paynter G, Witten I, Nevill-Manning C and Frank E 1999 Improving browsing in digital libraries with keyphrase indexes. *Decision Support Systems* **27**(1–2), 81–104.

Hulth A 2003 Improved automatic keyword extraction given more linguistic knowledge. *Proceedings of the 2003 Conference on Empirical Methods in Natural Language Processing*, vol. 10, pp. 216–223 Association for Computational Linguistics, Morristown, NJ, USA.

Hulth A 2004 *Combining machine learning and natural language processing for automatic keyword extraction*. Stockholm University, Faculty of Social Sciences, Department of Computer and Systems Sciences (together with KTH).

Jones K 1972 A statistical interpretation of term specificity and its application in retrieval. *Journal of Documentation* **28**(1), 11–21.

Jones S and Paynter G 2002 Automatic extraction of document keyphrases for use in digital libraries: evaluation and applications. *Journal of the American Society for Information Science and Technology*.

Matsuo Y and Ishizuka M 2004 Keyword extraction from a single document using word co-occurrence statistical information. *International Journal on Artificial Intelligence Tools* **13**(1), 157–169.

Mihalcea R and Tarau P 2004 Textrank: Bringing order into texts. In *Proceedings of EMNLP 2004* (ed. Lin D and Wu D), pp. 404–411. Association for Computational Linguistics, Barcelona, Spain.

Salton G, Wong A and Yang C 1975 A vector space model for automatic indexing. *Communications of the ACM* **18**(11), 613–620.

Whitney P, Engel D and Cramer N 2009 Mining for surprise events within text streams. *Proceedings of the Ninth SIAM International Conference on Data Mining*, pp. 617–627. Society for Industrial and Applied Mathematics.

2

Algebraic techniques for multilingual document clustering

Brett W. Bader and Peter A. Chew

2.1 Introduction

Pages on the World Wide Web have tremendous variation, covering a wide range of topics and viewpoints. Some are news pages, others are blogs. Given the sheer volume of documents on the Web, clustering these pages by topic would be a challenging problem. But web pages could be in any language, which complicates an already challenging text mining problem.

In a series of articles published largely in the computational linguistics literature, we have outlined a number of computational techniques for clustering documents in a multilingual corpus. This chapter reviews these techniques, provides some additional insight into these techniques, and presents some recent advances. Specifically, we show multiple algebraic models for this problem that were developed recently and that use matrix and tensor manipulations. These methods can be applied not just to pairs of languages, but also to groups of languages when a suitable multi-parallel corpus exists (Chew and Abdelali 2007).

In Sections 2.2 and 2.3, we review the problem and our experimental setup for multilingual document clustering. Then, in Sections 2.4–2.9 we present our

Text Mining: Applications and Theory edited by Michael W. Berry and Jacob Kogan
© 2010, John Wiley & Sons, Ltd

various approaches and their results. Section 2.10 discusses our results and summarizes our contribution.

2.2 Background

An early approach for dealing with documents in an information retrieval (IR) setting was the vector space model (VSM) of Salton (Salton 1968; Salton and McGill 1983). The principle behind the VSM is that a vector, with elements representing individual terms, may encode a document's meaning according to the relative weights of these term elements. Then one may encode a corpus of documents as a term-by-document matrix X of column vectors such that the rows represent terms and the columns represent documents. Each element x_{ij} tabulates the number of times term i occurs in document j. This matrix is sparse due to the Zipfian distribution of terms in a language (Zipf 1935).

As a practical matter for better performance, the term counts in X often are scaled. Many scaling approaches have been proposed, but the two most popular, based on their widespread availability in software such as SAS, are TFIDF (Term Frequency Inverse Document Frequency) and log-entropy scaling. Other approaches have been considered by Chisholm and Kolda (1999). We consider only the log-entropy scaling (see Equation (2.2)) in our approach here.

In 1990, Deerwester et al. (1990) proposed analyzing term-by-document matrices using the singular value decomposition (SVD) to organize terms and documents into a common semantic space based upon term co-occurrence. Because the approach claimed to organize the surface terms into their underlying semantics, the approach became known as latent semantic analysis (LSA).

In LSA a singular value decomposition of the (scaled) term–document matrix X is computed

$$X = USV^{T}. \tag{2.1}$$

Typically, a truncated SVD is computed such that a small number of columns (relative to the overall size of X) are retained. This amounts to keeping just the first R singular values in S (and correspondingly the first R columns of U and V). This low-rank approximation to X is in effect a dimensionality reduction that retains the most important information and leaves out noisier information. Projecting documents into this smaller dimensional subspace, one obtains feature vectors that may be used for similarity calculations or machine-learning tasks (e.g. (Chew et al. 2008a)).

As a statistics-based approach rooted in linear algebra and matrix computations, LSA has spawned many variations and new application areas. Pertaining to our current problem, Landauer and Littman (1990) extended latent semantic indexing by using a collection of abstracts in more than one language (English and French). Each 'document' is treated as the combination (in the bag of words sense) of French and English versions of the same abstract, and a multilingual space from LSA consists of terms from both languages coupled together. Their experiments showed that the two-language space was better for cross-language retrieval than single-language spaces. Queries in one language for retrieval in another language were shown to be just as effective as first translating the query into the language of a monolingual corpus. Young (1994) also uses only two languages (Greek and English), and the source data was the Gospels. He shows that LSA is effective in retrieving documents from either language without having to translate the user's query. The aspect that differentiates these studies from our work is that we consider more than just pairs of languages for cross-language information retrieval.

2.3 Experimental setup

For multilingual information retrieval experiments, one needs a multi-parallel corpus, which means that each document has a complete translation in all the languages. While many multilingual corpora exist and would work well, we use the Bible and Quran as our multilingual corpora. Both are carefully translated and are manually parallel aligned at the verse level (each verse contains roughly a sentence or two). Such fine-grained parallelism helps our machine-learning techniques learn concepts from word co-occurrences.

For training and testing purposes, we limited the selection of languages to Arabic, English, French, Russian, and Spanish. The lexical statistics of these translations of the Bible are listed in Table 2.1. The linguistic differences among the languages are evident in the table. English has the fewest unique terms, whereas Arabic has nearly five times as many unique terms and just over half as many total words for the whole translation. The ordering in Table 2.1 roughly corresponds to the ordering of languages on a spectrum that linguists identify on one end as 'isolating' (one morpheme, or individual unit of meaning, per word) and on the other end as 'synthetic' (high morpheme-per-word ratio).

English is largely an isolating language because most words have one or just a few morphemes. For example, verbs may have markers for tense (e.g. the morpheme 'ed' is the past tense inflection); nouns may be compound or plural (e.g. the morpheme 's' often indicates a plural noun).

German is closer to the other end of the spectrum as a synthetic language because it has many compound nouns composed of individual morphemes. But

Table 2.1   Lexical statistics of the translations of the Bible used for training.

Language (translation)	Unique terms	Total word count
English (King James)	12 335	789 744
French (Darby)	20 428	812 947
Spanish (Reina Valera 1909)	28 456	704 004
Russian (Synodal 1876)	47 226	560 524
Arabic (Smith Van Dyke)	55 300	440 435

there are other languages with even starker differences. Payne (Payne 1997) cites an illustrative example that comes from Yup'ik Eskimo, *tuntussuqatarnik-saitengqiggtuq*, which means 'he had not yet said again that he was going to hunt reindeer'. This word is composed of many morphemes, as evidenced by the fact that the English translation has multiple words. For example, the first morpheme, *tuntu*, refers to reindeer. So if the concept were to change instead to 'she was going to hunt reindeer', then there would be a whole new unique word starting with *tuntu* containing only some of the morphemes from the example along with a different morpheme due to the change in gender of the subject. Thus, it is easy to see why such a language would prove troublesome for VSMs. Each word, which is packed with more meaning, is represented by a single direction in vector space instead of a collection of directions based on its constituent morphemes.

Hence, these language differences provide a challenge to statistical techniques that rely on co-occurrence patterns. Synthetic languages, which have more unique terms representing more diverse concepts, will have fewer terms co-occurring with other terms from an isolating language, making it more difficult to learn from relationships from co-occurrence patterns.

For our system, we do not consider traditional stemming or stoplists because we want the most generalizable system that does not rely on expert knowledge of a language. We prefer to rely solely on the statistical properties of the corpus for an extensible system for languages that may be applied to less common or obscure languages.

The Bible has 31 226 verses, which we use as individual 'documents' in our training set. The Quran has 114 suras (or chapters), which we use as the documents in our test set. With the five languages, we have 570 individual test queries. For each new query document, we project its vector representation into the space of US^{-1} and compute a cosine similarity with all other document feature vectors. The highest similarity indicates the best match available, which for our case should be a matching translation of the query document. We use S^{-1} instead of other alternatives because if we consider the documents in X as our test set, then the projection of X on US^{-1} is close to the matrix V, which is the document-by-concept matrix from the SVD.

To assess the performance of our techniques, we consider two measures of precision used in multilingual IR. For the first, we split the test set into each of the 25 possible language-pair combinations, where these include each language to itself. For each pair, we have 228 distinct queries (i.e. chapters). The goal is to retrieve the corresponding translation of that chapter in the other language. We calculate the average precision at one document (P1), which is the average percentage of times that the translation of the query ranked highest. P1 may be calculated as an average over all queries for each language pair or as an overall average, which we report here. P1 is a fairly strict measure of precision that essentially measures success in retrieving documents when the source and target languages are specified.

For the second measure, we considered average multilingual precision at five documents (MP5), which is the average percentage of the top five documents that are translations of the query document. We calculate MP5 as an average for all queries and all languages. Essentially, MP5 measures success in multilingual clustering. MP5 is a stricter measure than P1; since the target language is not specified, there are more possibilities to choose from.

2.4 Multilingual LSA

In the context of cross-language IR, one starts with a parallel multilingual corpus. The approach used in Landauer and Littman (1990) and Young (1994) for pairs of languages, and used in Chew and Abdelali (2007) for multiple languages, is to stack all term–document matrices for each language, one on top of another; see Figure 2.1. The rows correspond to terms in all of the languages, and the truncated SVD finds the optimal rank R representation of this matrix. The factor matrices group terms and documents into orthogonal basis vectors based upon term and document co-occurrence patterns in X.

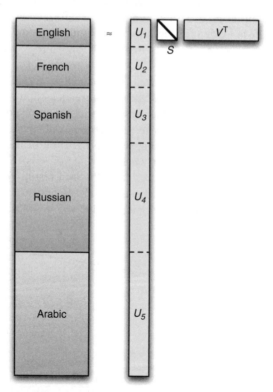

Figure 2.1 An illustration of the multilingual LSA using the SVD.

The best results for five languages and a rank 300 SVD give an average P1 score of 76.0% and an average MP5 score of 26.1%. While the P1 score is respectable, the MP5 score is disappointing because it means that documents are clustering more by language than by topic.

We observed in our results an imbalance in the importance of common terms (e.g. determiners, pronouns, conjunctions, prepositions) in the concept vectors of the U matrix. This fact stemmed from the way the standard log-entropy formula treated common terms with respect to other terms with higher information gain. This insight led us to modify the log-entropy formula so that the common terms with high entropy were less influential in the SVD. Our simple modification to log-entropy involved raising the global term weight to a power $\alpha > 1$:

$$X_{td} = \log(X_{td} + 1) \left[1 + \frac{H_t}{\log N} \right]^{\alpha} \tag{2.2}$$

where $H_t = \sum_d (X_{td}/F_t) \log(X_{td}/F_t)$ is the entropy of term t and F_t is the raw frequency of term t in the corpus.

The overall effect of this modification is that $\alpha > 1$ mitigates the influence of common terms in the SVD. As α increases, the 'weight' of elements in X shifts away from common terms to less common, more information-rich terms, and a corresponding shift is evident in the principal singular vectors. However, if α is too large, then the X matrix consists mainly of low-entropy terms (e.g. proper nouns).

Our computational studies showed that $\alpha = 1.8$ significantly improved retrieval results for all of our techniques. Figure 2.2 shows the global term weights for all terms in English. The first term index corresponds to the word 'and'. There are roughly 60 000 terms that appear only once each (so-called *hapax legomena*) in the Bible. These appear on the right of the plot and have a global term weight of one, no matter what the value of α is.

With the improved global term weighting, the best results for five languages and a rank 300 SVD give an average P1 score of 88.0% and an average MP5 score of 65.7%. We see a large increase in P1 (p value $= 7 \times 10^{-51}$) and a dramatic

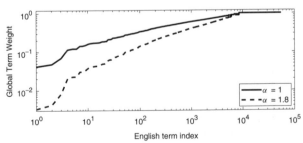

Figure 2.2 Improved term–document matrix weighting by raising global term weight to a power of α.

increase in multilingual precision (p value $= 0$). Nevertheless, the documents are still clustering more by language than by topic.

2.5   Tucker1 method

In Chew et al. (2007), we pursued a new paradigm in multilingual text analysis where, instead of stacking the language matrices one on top of another to create one tall matrix for the SVD, the matrices are stacked one behind another in a third dimension to form a multi-set array, see Figure 2.3.

   When the data is organized in this manner and all three dimensions are the same, the object is called an n-way array or a tensor, which we denote with a script font, e.g. $\mathcal{X}$. There are many decompositions or factorizations of tensors to choose from, several of which are generalizations of the matrix SVD (Kolda and Bader 2009). One of the most basic approaches to consider is the Tucker1 model (Kolda and Bader 2009; Tucker 1966), which finds a single orthonormal factor matrix in one of the modes that applies across all slices in parallel. Mathematically, the Tucker1 model is

$$X_k \approx A_k V^T \quad \text{for } k = 1, \ldots, K, \tag{2.3}$$

where the notation X_k and A_k refers to the kth frontal slice of tensors $\mathcal{X}$ and $\mathcal{A}$, respectively, with what is called slab notation. The matrix V is the set of principal eigenvectors of $\sum_k X_k^T X_k$ (which is the same as the principal right singular vectors of the matrix formed by stacking the slices X_k on top of each other). Each matrix A_k is the matrix that best fits the data in a least squares sense, which is just $A_k = X_k V$ because V is orthonormal.

   To use the same framework as outlined previously for multilingual LSA where we project new documents in the space of $U_k S_k^{-1}$, we normalize the columns in each A_k so that they have unit length and the weight is stored in a diagonal matrix S_k. Then the Tucker1 representation becomes

$$X_k \approx U_k S_k V^T \quad \text{for } k = 1, \ldots, K. \tag{2.4}$$

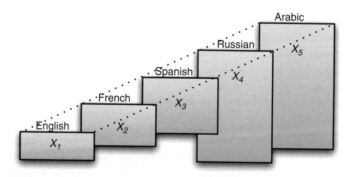

Figure 2.3   Multi-set array of term-by-document matrices.

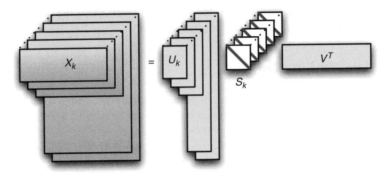

Figure 2.4   An illustration of the Tucker1 model.

For our case where the row dimension is not constant, however, we may assume that the tensor has a row dimension of the largest matrix and that the other smaller matrices are padded with rows of zeros in order to adapt the Tucker1 model. The resulting factor matrices U_k will have a corresponding number of zero rows. Figure 2.4 shows the Tucker1 model.

Using a rank 300 Tucker1 model, we get an average P1 score of 89.5% and an average MP5 score of 71.3%. With this tensor representation, we see a small increase over SVD in P1 (p value $= 8 \times 10^{-3}$) and a large increase in multilingual precision (p value $= 4 \times 10^{-11}$). However, the fact that each U_k does not form an orthogonal space in the Tucker1 model may be limiting the performance of this tensor approach. When projecting new documents onto these oblique axes to get document feature vectors, distances between features are distorted, which could adversely affect cosine similarity calculations.

2.6   PARAFAC2 method

PARAFAC2 (Harshman 1972) is a tensor decomposition that has orthogonal basis vectors and is extensible to multi-set data. PARAFAC2 has been used in the analysis of chemometric data, specifically in chromatography with retention time shifts among samples. In chromatography, each sample being analyzed may have a different elution profile, meaning that the signal may take longer or shorter to collect. In such cases, each matrix may have a different number of rows, which is just like the form of our multilingual multi-set data.

In Chew et al. (2007) we apply the PARAFAC2 technique to the multi-set term–document array of Figure 2.3. The mathematical model of PARAFAC2 is

$$X_k \approx U_k H S_k V^T \quad \text{for } k = 1, \ldots, K, \tag{2.5}$$

where each U_k is an orthogonal matrix that may have a different number of rows for each k, H is a dense matrix that is predominantly diagonal for our application, S_k is a diagonal matrix containing weights for each level of k, and

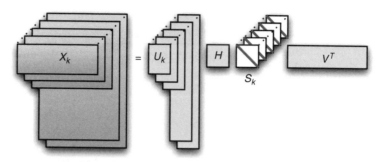

Figure 2.5   An illustration of the PARAFAC2 model.

V is a dense matrix that is not necessarily orthogonal. Figure 2.5 shows the PARAFAC2 model. We project new documents in the space of $U_k S_k^{-1}$.

The algorithm to fit a PARAFAC2 model is decidedly more complex than for Tucker1, so we will only refer to the algorithm in Kiers et al. (1999), which we implemented in MATLAB using the Tensor Toolbox (Bader and Kolda 2006, 2007a,b).

Due to memory constraints, we were not able to compute a rank 300 PARAFAC2 model. Instead we computed a rank 240 PARAFAC2 model, which provided an average P1 score of 89.8% and an average MP5 score of 78.5%. With this tensor representation, and even though the rank of the model is lower than previously, we see a large and highly significant increase in MP5 over Tucker1 (p value $= 2 \times 10^{-17}$). However, the increase over Tucker1 is insignificant for P1 (p value $= 0.6$).

2.7   LSA with term alignments

In Bader and Chew (2008) we returned to the matrix formulation of the term-by-document matrix. Our approach was inspired by Hendrickson (2007), who showed that LSA was related to Fiedler vectors of a graph Laplacian. This connection suggested a means to incorporate additional information beyond just term–document relationships into the SVD.

The basis of this approach is that the SVD may be calculated in several different ways; see Table 2.2. If we consider the third option listed, where one can get U and V from the eigenvectors of a block matrix with X and X^T on the off diagonal, then we may add information to the diagonal blocks that complements the information only found in X. In the context of LSA and a term–document matrix X, these diagonal blocks correspond to term–term and document–document similarity information. In Bader and Chew (2008) we consider adding information only to the first diagonal block (labeled D_1) corresponding to the terms; see Figure 2.6.

There are several possible methods which can be used to add information to D_1. Conceptually, the simplest approach involves consulting a dictionary

Table 2.2 Calculating the SVD $X = U\Sigma V^T$ may be accomplished via an eigendecomposition of different matrices involving X.

Matrix	Eigenvectors		Eigenvalues
XX^T $\rightarrow$ U		&	Σ^2
X^TX $\rightarrow$ V		&	Σ^2
$\begin{bmatrix} 0 & X \\ X^T & 0 \end{bmatrix} \rightarrow$	$\frac{1}{\sqrt{2}}\begin{pmatrix} U_+ & \sqrt{2}U_0 & -U_+ \\ V & 0 & V \end{pmatrix}$	&	$\begin{pmatrix} \Sigma & & \\ & 0 & \\ & & -\Sigma \end{pmatrix}$

U_+ is the matrix of singular vectors with positive singular values, U_0 is the matrix of singular vectors with zero singular values.

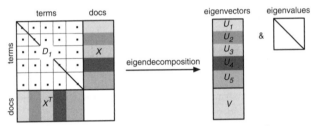

Figure 2.6 Eigendecomposition of block matrix with term-alignment information yields stronger cross-lingual term relationships.

and populating the block so that $D_{ij} = 1$ if the term pair (i, j) occurs in a dictionary, and zero otherwise. Another option, which was used in Bader and Chew (2008), involves computing the pairwise mutual information (PMI) of two terms appearing together in the same documents across the whole corpus. This draws upon one of the ideas which underpins statistical machine translation (SMT) (Brown et al. 1994). To preserve sparsity in the matrix, we retain the value only for the pair (i, j) that has the highest PMI in both directions. Because the resulting matrix is not symmetric and symmetry is needed in D_1 to obtain real eigenvalues, we symmetrize the matrix using a modified Sinkhorn balancing procedure. Sinkhorn balancing (Sinkhorn 1964) is also needed to equalize contributions between terms. The standard Sinkhorn balancing procedure normalizes the row and column sums to one, but we use a modified procedure that makes each row and column of D_1 have unit length. This modification was found to produce better results than creating a doubly stochastic matrix D_1. All together, we call this technique LSA with term alignments (LSATA).

By adding term-alignment information to the diagonal block, we strengthen the co-occurrence information that LSA normally finds in the parallel corpus via the SVD. To understand this mathematically, we consider the solution obtained from LSA and then apply a power method to update U and V. Here is one iteration of the power method on our block matrix:

$$U_{new} = D_1 U + XV, \qquad (2.6)$$

$$V_{new} = X^T U. \qquad (2.7)$$

The terms XV and $X^T U$ are the standard relationships in LSA. The term $D_1 U$ is new, and it reinforces term–term relationships from external information (although note that under our approach, the information is not 'external' in that it is implied by the same corpus from which we get the term-by-document matrix). A graphical representation of this interpretation is shown in Figure 2.7, where in one of the concept vectors in U the term 'house' dominates the corresponding terms in Spanish and French, for example. After multiplication with D_1, the relationship between these three words is strengthened, and all three terms have similar values.

This observation leads to another consideration: the weighting of D_1 relative to X. If the values of D_1 are very small compared to X, then any contribution from D_1 will be negligible. The opposite happens if the values in D_1 are very large compared to X. Hence, the matrices D_1 and X must be numerically balanced by, say, multiplying D_1 by some parameter β. For our corpus and particular scaling of D_1 (Sinkhorn-balanced PMI) and X (log-entropy with $\alpha = 1.8$), we determined empirically that a value of $\beta = 12$ provides good results. Alternatively, β can be determined automatically by routinely balancing the two contributions from $D_1 U$ and XV in Equations (2.6)–(2.7). One possible approach is to set

$$\beta = \frac{\|XV\|_F}{\|D_1 U\|_F}. \qquad (2.8)$$

Algorithmically, β could be computed iteratively inside an eigensolver or externally by looping over an eigensolver and adjusting β until it converges to a constant value.

In our numerical experiments, using a rank 300 LSATA model and $\beta = 12$, we get an average P1 score of 91.8% and an average MP5 score of 80.7%. With this matrix representation, we see a small increase over PARAFAC2 in P1 and in

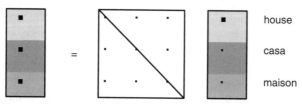

Figure 2.7   Term-alignment matrix D_1 strengthens the cross-lingual term relationships in U identified by LSA.

multilingual precision. Although these increases are small, they are statistically significant (p values of 1×10^{-4} and 4×10^{-3}, respectively).

2.8   Latent morpho-semantic analysis (LMSA)

In Chew et al. (2008b) we investigated alternate formulations of cross-language retrieval using the VSM. A recurring pattern in our results was that Arabic and Russian tended to have lower P1 scores than English, French, and Spanish. This occurred irrespective of whether Arabic and Russian were the source or target languages of the query, and it held for all of our techniques. This pattern suggested a linguistic explanation to the problem and a corresponding linguistic solution.

As discussed previously, languages fall on a spectrum from isolating to synthetic. Arabic and Russian are synthetic languages, where meaning is shaped through inflection and suffixation of words. This means that these languages have more unique terms (see Table 2.1), which means that, for example in English, the terms *walk*, *walks*, *walking*, and *walked* would correspond to separate rows in X, and the co-occurrence patterns of these words may indicate that they are not related. For morphologically complex languages like Arabic and Russian, even more extreme examples could be found.

To address this problem, we have developed a morphologically more sophisticated alternative to LSA, which we call latent morpho-semantic analysis (LMSA) (Chew et al. 2008b). In this technique we perform a statistical analysis of the language to identify tokenizations of character n-grams that maximize mutual information among all possible nonoverlapping n-grams. We then use these tokenizations to form a morpheme-by-document matrix instead of a term-by-document matrix, weight it using log-entropy scaling (or other), and apply the SVD to get a morpheme-by-concept matrix U and corresponding singular values S, which we subsequently use in the standard way.

The benefits of this approach are twofold. First, all of the benefits of LSA (language independence, speed of implementation, fast runtime processing) are retained in this method. Second, we are more able to deal with out-of-vocabulary terms in new documents because they may be broken down into their constituent morphemes, which are more likely to be represented in the training corpus. Our approach is related to stemming, except that all parts of the word are retained in a morpheme-by-document matrix, not just the stems. Furthermore, this procedure may be performed by a statistical analysis of the language, so no language expertise is required.

Using a rank 300 LMSA model, we get an average P1 score of 88.7% and an average MP5 score of 73.7%. With this linguistic representation, we see performance about on par with the Tucker1 tensor technique: no statistical difference in P1 but a small (2.4%) improvement in multilingual precision (p value $= 5 \times 10^{-3}$).

2.9   LMSA with term alignments

With the development of LMSA, it is a simple extension to consider the term-alignment framework of LSATA using morphemes instead of terms. We may call this technique LMSATA, or LMSA with term alignments (in this case, terms refer to morphemes). Morpheme alignments are determined using mutual information in the same manner as terms are with LSATA. Then a 2×2 block matrix is formed with morpheme alignments in D_1 (scaling factor $\beta = 12$) and morpheme-by-document matrix X with log-entropy weighting. An eigendecomposition of this matrix yields eigenvectors from which we extract individual U matrices for each language.

Using a rank 300 LMSATA model, we get an average P1 score of 94.6% and an average MP5 score of 81.7%. With this linguistic representation, we see an increase over the previous best method, LSATA, with a large gain in P1 (p value $= 5.1 \times 10^{-8}$) and a slight but insignificant increase for MP5 (p value $= 0.18$).

2.10   Discussion of results and techniques

The results from all of our methods are tabulated in Table 2.3. Note that a number of techniques from standard IR and computational linguistics were combined to achieve much higher multilingual precisions. In fact, our best method reported here, LMSATA, relies on a broad collection of techniques including: (1) morphological analysis of language using techniques from statistical machine translation; (2) techniques from latent semantic analysis, including dimensionality reduction using the SVD; and (3) numerical linear algebra for simultaneously analyzing term co-occurrences and term–term alignments.

It is difficult to compare these techniques in terms of computational performance because they were not implemented on a single machine/architecture; some are parallel codes, others are serial MATLAB implementations. Generally speaking, the SVD-based techniques, such as LSA and LMSA, are the fastest.

Table 2.3   Aggregate results from all algebraic techniques.

Method	Average P1	Average MP5
SVD/LSA ($\alpha = 1.0$)	76.0%	26.1%
SVD/LSA ($\alpha = 1.8$)	88.0%	65.7%
Tucker1	89.5%	71.3%
PARAFAC2	89.8%	78.5%
LSATA	91.8%	80.7%
LMSA	88.7%	73.7%
LMSATA	94.6%	81.7%

The eigenvector-based approach of LSATA and LMSATA requires more time due to the larger matrix and term-alignment step. The tensor-based techniques Tucker1 and PARAFAC2 are the slowest due to the data being organized as a large three-way array. The morphological tokenization of LMSA and LMSATA adds an extra processing step that adds time to both the training and test sets, and the resulting morpheme-by-document matrix is smaller yet denser.

As a demonstration of what is possible when the framework achieves high multilingual precision (around 90%), we present in Figures 2.8 and 2.9 a visualization of how the books of the Bible, color-coded according to language, are represented in two-dimensional space. Note that the books cluster first to their counterparts in other languages, and then into larger clusters containing related books. In particular, Figure 2.9 shows that John and Acts have tight clusters, while there is some mixing among Matthew, Mark, and Luke, which seems reasonable; Bible scholars call these three synoptic gospels because they share a similar perspective. This kind of visualization is possible only when

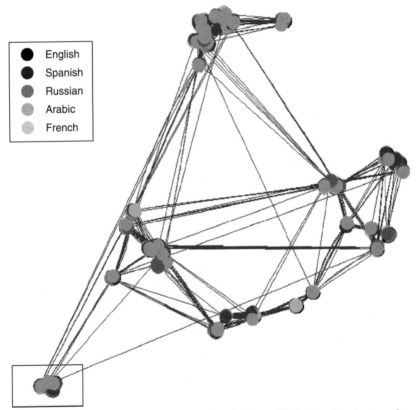

Figure 2.8   Visualization of clustering of multilingual Bible books. Rectangle represents area of detail shown in Figure 2.9.

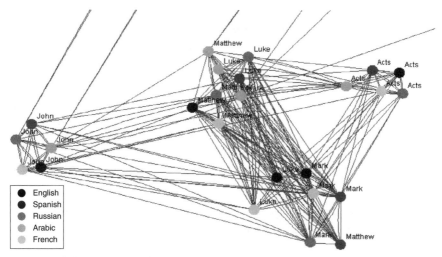

Figure 2.9   Partial visualization of clustering of multilingual Bible books.

multilingual precision is satisfactorily high. In summary, these techniques have effectively allowed us to factor out language, focusing only on topic, just as we had hoped.

2.11   Acknowledgements

Sandia is a multiprogram laboratory operated by Sandia Corporation, a Lockheed Martin Company, for the United States Department of Energy's National Nuclear Security Administration under contract DE-AC04-94AL85000.

References

Bader BW and Chew PA 2008 Enhancing multilingual latent semantic analysis with term alignment information. *COLING 2008*.

Bader BW and Kolda TG 2006 Algorithm 862: MATLAB tensor classes for fast algorithm prototyping. *ACM Transactions on Mathematical Software* **32**(4), 635–653.

Bader BW and Kolda TG 2007a Efficient MATLAB computations with sparse and factored tensors. *SIAM Journal on Scientific Computing* **30**(1), 205–231.

Bader BW and Kolda TG 2007b Tensor toolbox for MATLAB, version 2.2. http://csmr.ca.sandia.gov/~tgkolda/TensorToolbox/.

Brown PF, Della Pietra VJ, Della Pietra SA and Mercer RL 1994 The mathematics of statistical machine translation: Parameter estimation. *Computational Linguistics* **19**(2), 263–311.

Chew P and Abdelali A 2007 Benefits of the massively parallel Rosetta Stone: Cross-language information retrieval with over 30 languages. *Proceedings of the Association for Computational Linguistics*, pp. 872–879.

Chew P, Kegelmeyer P, Bader B and Abdelali A 2008a The knowledge of good and evil: Multilingual ideology classification with PARAFAC2 and machine learning. *Language Forum* **34**(1), 37–52.

Chew PA, Bader BW and Abdelali A 2008b Latent morpho-semantic analysis: Multilingual information retrieval with character n-grams and mutual information. *COLING 2008*.

Chew PA, Bader BW, Kolda TG and Abdelali A 2007 Cross-language information retrieval using PARAFAC2. *KDD'07: Proceedings of the 13th ACM SIGKDD International Conference on Knowledge Discovery and Data Mining*, pp. 143–152. ACM Press, New York.

Chisholm E and Kolda TG 1999 New term weighting formulas for the vector space method in information retrieval. Technical Report ORNL-TM-13756, Oak Ridge National Laboratory, Oak Ridge, TN.

Deerwester SC, Dumais ST, Landauer TK, Furnas GW and Harshman RA 1990 Indexing by latent semantic analysis. *Journal of the American Society for Information Science* **41**(6), 391–407.

Harshman RA 1972 PARAFAC2: Mathematical and technical notes. *UCLA Working Papers in Phonetics* **22**, 30–47.

Hendrickson B 2007 Latent semantic analysis and Fiedler retrieval. *Linear Algebra and its Applications* **421**(2–3), 345–355.

Kiers HAL, Ten Berge JMF and Bro R 1999 PARAFAC2 – Part I. A direct fitting algorithm for the PARAFAC2 model. *Journal of Chemometrics* **13**(3–4), 275–294.

Kolda TG and Bader BW 2009 Tensor decompositions and applications. *SIAM Review* **15**(3), 455–500.

Landauer TK and Littman ML 1990 Fully automatic cross-language document retrieval using latent semantic indexing. *Proceedings of the 6th Annual Conference of the UW Centre for the New Oxford English Dictionary and Text Research*, pp. 31–38, UW Centre for the New OED and Text Research, Waterloo, Ontario.

Payne TE 1997 *Describing Morphosyntax: A guide for field linguists*. Cambridge University Press, Cambridge, UK.

Salton G 1968 *Automatic Information Organization and Retrieval*. McGraw-Hill, New York.

Salton G and McGill M 1983 *Introduction to Modern Information Retrieval*. McGraw-Hill, New York.

Sinkhorn R 1964 A relation between arbitrary positive matrices and doubly stochastic matrices. *Annals of Mathematical Statistics* **35**(2), 876–879.

Tucker LR 1966 Some mathematical notes on three-mode factor analysis. *Psychometrika* **31**, 279–311.

Young P 1994 *Cross language information retrieval using latent semantic indexing*. Master's thesis University of Knoxville Knoxville, TN.

Zipf GK 1935 *The Psychobiology of Language*. Houghton-Mifflin, Boston, MA.

3

Content-based spam email classification using machine-learning algorithms

Eric P. Jiang

3.1 Introduction

With the rapid growth of the Internet and advances in computer technology email has become a preferred form of communication and information exchange for both business and personal purposes. It is fast and convenient. In recent years, however, the effectiveness and confidence in email have been diminished quite noticeably by spam email, or bulk unsolicited and unwanted email messages. Spam email has been a painful annoyance for email users with an overwhelming amount of unwelcome messages flowing into their mailboxes. Now, it has also evolved into a primary medium for spreading phishing scams and malicious viruses. The cost of spam in the United States alone in terms of decreased productivity and increased technical expenses for businesses has reached tens of billions of dollars annually.[1] Worldwide spam volume has increased significantly and during the first quarter of 2008, spam email accounted for more than nine out of every ten email messages sent over the Internet.[2]

---

[1] http://www.spamlaws.com/spam-stats.html
[2] http://www.net-security.org/

Text Mining: Applications and Theory   edited by Michael W. Berry and Jacob Kogan
© 2010, John Wiley & Sons, Ltd

Over the years, various spam filtering technology and anti-spam software products have been developed and deployed. Some of them are designed to detect and stop spam email at the TCP/IP or SMTP level and may rely on DNS blacklists of domain names that are known to originate spam. This approach has been commonly used. However, it can be insufficient due to the lack of accuracy of the name lists, since spammers can now register hundreds of free webmail services such as Hotmail and Gmail and then rotate them every few minutes during a spam campaign. The other major type of spam filtering technology functions at the client level. Once an email message is downloaded, its content can be examined to determine whether the message is spam or legitimate. Several supervised machine-learning algorithms have been used in client-side spam detection and filtering. Among them, naive Bayes (Mitchell 1997; Sahami et al. 1998), boosting algorithms such as logitBoost (Androutsopoulos et al. 2004; Friedman et al. 2000), support vector machines (SVMs) (Christianini and Shawe-Taylor 2000; Drucker et al. 1999), instance-based algorithms such as k-nearest neighbor (Aha and Albert 1991), and Rocchio's classifier (Rocchio 1997) are commonly cited. More recently, a number of other interesting algorithms for spam filtering have been developed. One uses an augmented latent semantic indexing (LSI) space model (Jiang 2006) and another applies a radial basis function (RBF) neural network (Jiang 2007).

This chapter considers five supervised machine-learning algorithms for an evaluation study of spam filtering application. The algorithms selected in this study include widely used ones with good classification results and some recently proposed methods. More specifically, we evaluate these five classification algorithms: naive Bayes classifier (NB), support vector machines (SVMs), logitBoost algorithm (LB), augmented latent semantic indexing space model (LSI) and radial basis function (RBF) networks.

Spam filtering is a cost-sensitive classification task since misclassifying legitimate email (a *false positive* error) is generally more costly than misclassifying spam email (a *false negative* error). Fairly recently, there have been several studies (Androutsopoulos et al. 2004; Zhang et al. 2004) surveying machine-learning techniques in spam filtering. Using a constant λ to measure the higher cost of false positives, these studies have evaluated several algorithms on spam filtering by integrating the λ value or a function of λ into the algorithms through a variety of cost-sensitive adjustment strategies. This was done by increasing algorithm thresholds on spam confidence scores, adding more weights on legitimate training samples, or empirically adjusting algorithm decision thresholds using cross-validation. Different adjustment strategies have also been applied to different algorithms in the studies. Since all the algorithms were designed with cost-insensitive tasks in mind, applying such simple cost-sensitive adjustments on the algorithms can produce unreliable results. Apparently this insufficiency has been recognized and, for some algorithms, the studies reported only the best results among several adjustment trials.

This chapter provides a related study of five machine-learning algorithms on spam filtering from a different perspective. The main objective of the study is to learn whether and to what extent the algorithms are adaptable and applicable to the cost-sensitive email classification problem and to identify the characteristics of the algorithms most suitable for adaptability. In this study, we selected two benchmark email testing corpora for experiments that were constructed from two different languages and have reverse ratios of the number of spam emails to the number of legitimate emails in the training data. We also vary feature size in the experiments to analyze the usefulness of feature selection for these algorithms.

The rest of the chapter is organized as follows. In Section 3.2, the five machine-learning algorithms that are investigated for spam filtering applications are briefly described. In Section 3.3, several data preprocessing procedures, including feature selection and message representation, are discussed. Spam filtering is a cost-sensitive classification task and a related discussion of effectiveness measures is included in Section 3.4. We then compare the algorithms, using two popular email testing corpora. The experimental results and analysis are reported in Section 3.5, and an empirical comparison of the characteristics of the five classifiers is presented in Section 3.6. Finally, some concluding remarks are provided in Section 3.7.

3.2   Machine-learning algorithms

Spam email filtering is an application of automated text classification with two categories. A number of machine-learning algorithms, which have been successfully used in text classification (Sebastiani 2002), can also be applied in spam filtering. Given a collection of labeled email samples, these algorithms can learn from the samples to classify previously unseen email into the categories based on their content. The algorithms of NB, LB, SVM, augmented LSI, and RBF are among those that have achieved good performance for spam filtering. They are included in this study and are briefly described in this section.

In this chapter, we use $D = \{d_1, d_2, \ldots, d_n\}$ to denote a training set of email samples with size n and $C = \{c_l, c_s\}$ the email categories (c_l, legitimate; c_s, spam). We assume each email message d_i can be expressed as a numeric vector representing the weights of terms or features $d_i = (t_1, t_2, \ldots, t_m) \in \Re^n$ (see Section 3.3.2).

3.2.1   Naive Bayes

The NB classifier is a probabilistic learning algorithm that derives from Bayesian decision theory (Mitchell 1997). The probability of a message d being in class c, $P(c|d)$, is computed as

$$P(c|d) \propto P(c) \prod_{k=1}^{m} P(t_k|c), \qquad (3.1)$$

where $P(t_k|c)$ is the conditional probability of feature t_k occurring in a message of class c and $P(c)$ is the prior probability of a message occurring in class c.

$P(t_k|c)$ can be used to measure how much evidence t_k contributes that c is the correct class (Manning et al. 2008). In email classification, the class of a message is determined by finding the most likely or maximum a posteriori (MAP) class c_{MAP} defined by

$$c_{MAP} = \arg \max_{c \in \{c_l, c_s\}} P(c|d) = \arg \max_{c \in \{c_l, c_s\}} P(c) \prod_{k=1}^{m} P(t_k|c). \qquad (3.2)$$

Since Equation (3.2) involves a multiplication of many conditional probabilities, one for each feature, the computation can result in a floating point underflow. In practice, the multiplication of probabilities is often converted to an addition of logarithms of probabilities and, therefore, the maximization of the equation is alternatively performed by

$$c_{MAP} = \arg \max_{c \in \{c_l, c_s\}} \left[\log P(c) + \sum_{k=1}^{m} \log P(t_k|c) \right]. \qquad (3.3)$$

All model parameters, i.e. class priors and feature probability distributions, can be estimated with relative frequencies from the training set D. Note that when a given class and message feature do not occur together in the training set, the corresponding frequency-based probability estimate will be zero, which would make the right hand side of Equation (3.3) undefined. This problem can be mitigated by incorporating some correction such as Laplace smoothing in all probability estimates.

NB is a simple probability learning model and can be implemented very efficiently with a linear complexity. It applies a simplistic or naive assumption that the presence or absence of a feature in a class is completely independent of any other features. Despite the fact that this oversimplified assumption is often inaccurate (in particular for text domain problems), NB is one of the most widely used classifiers and possesses several properties (Zhang 2004) that make it surprisingly useful and accurate.

3.2.2   LogitBoost

LB is a boosting algorithm that implements forward stagewise modeling to form additive logistic regression (Friedman et al. 2000). Like other boosting methods, LB adds base models or learners of the same type iteratively, and the construction of each new model is influenced by the performance of those preceding ones. This is accomplished by assigning weights to all training samples and adaptively updating the weights through iterations. Suppose f_m is the mth base learner and $f_m(d)$ is the prediction value of message d. After f_m is constructed and added to the ensemble, the weights on training samples are updated in such a way that the subsequent base learner f_{m+1} will focus more on those difficult samples to classify by f_m. In the iteration process, the probability of d being in class c

is estimated by applying a sigmoid function, which is also known as the *logit* transformation, to the response of the ensemble that has been built so far, i.e.

$$P(c|d) = \frac{e^{F(d)}}{1 + e^{F(d)}}, \quad F(d) = \frac{1}{2}\sum f_m(d). \tag{3.4}$$

Once the iteration terminates and the final ensemble F is created, the classification of target email messages is determined by the probability in Equation (3.4).

A popular base learner choice for LB is decision stump, a one-level decision tree that uses an attribute in training data to classify training samples into categories. In text classification, since we deal with continuous attributes, the decision tree is actually a threshold function on one of the data attributes and hence it becomes a regression stump (Androutsopoulos et al. 2004). It can be shown that the LB algorithm maximizes the probability of the data with respect to the ensemble if each base learner f_m is determined by minimizing the squared error on the fitted regression of weighted training data (Witten and Frank 2005). The model's iteration number m is specified by the user and we set it to 50, which is the smallest feature size used in this study.

3.2.3   Support vector machines

SVMs (Christianini and Shawe-Taylor 2000) have been considered the most promising algorithm in text classification. The algorithm uses linear models to implement nonlinear category boundaries by transforming a given instance space into a linearly separable one through nonlinear mappings. In the transformed space, an SVM constructs a separating hyperplane that maximizes the distance between the training samples of two categories. This is done by selecting two parallel hyperplanes that are each tangent to at least one sample of its category; such samples on the tangential hyperplanes are called the support vectors. The distance between the two tangential planes is the margin of the classifier, which is to be maximized, and that is why a linear SVM is also known as a maximal margin classifier.

Assume the class variable for the ith training sample is $c_i = \{1, -1\}$, indicating the spam (1) or legitimate (-1) category, respectively. A hyperplane in the sample space can be written as

$$w \cdot d + b = 0, \tag{3.5}$$

where w is a normal vector that is perpendicular to the hyperplane, and b is a bias term. If the given training data is linearly separable, we can select two hyperplanes that contain no points between them and then maximize the distance (margin) between the hyperplanes, which is $2/\|w\|$. Maximizing the margin is equivalent to solving the following constrained minimization problem:

$$\min_{w} \frac{\|w\|^2}{2}, \quad \text{subject to } c_i(w \cdot d_i + b) \geq 1. \tag{3.6}$$

The optimization problem in Equation (3.6) can be solved by the standard Lagrange multiplier method with the new objective function:

$$\frac{\|w\|^2}{2} - \sum_i \lambda_i [c_i(w \cdot d_i + b) - 1]. \tag{3.7}$$

Since the Lagrangian involves a large number of parameters, this is still a difficult problem. Fortunately, the problem can be simplified by transforming the Lagrangian in Equation (3.7) into the following dual formation that contains only Lagrange multipliers:

$$\max \sum_i \lambda_i - \frac{1}{2} \sum_{i,j} \lambda_i \lambda_j c_i c_j d_i \cdot d_j, \text{ subject to } \lambda_i \geq 0, \text{ and } \sum_i \lambda_i c_i = 0. \tag{3.8}$$

The dual optimization problem can usually be solved by using some numerical quadratic programming techniques such as the sequential minimal optimization algorithm (Platt 1999). The terms λ_i from Equation (3.8) are used to define the decision boundary

$$\left(\sum_i \lambda_i c_i d_i \cdot d \right) + b = 0. \tag{3.9}$$

In order to deal with the cases where the training samples cannot be fully separated and also small misclassification errors are permitted, the so-called *soft margin method* was developed for choosing a hyperplane that intends to reduce the number of errors committed by the decision boundary while maximizing the width of the margin. The method introduces a positive-valued slack variable ξ that measures the degree of misclassification error on a sample and solves the following modified optimization problem:

$$\min_w \frac{\|w\|^2}{2} + C \sum_i \xi_i, \text{ subject to } c_i(w \cdot d_i + b) \geq 1 - \xi_i, \tag{3.10}$$

where a linear penalty function is used and C is a user-specified constant that determines an error tolerance level. In our experiments, we set $C = 1$.

The linear SVM described above can be extended into a nonlinear classifier. Conceptually, we could just transform the training data (where no linear decision boundaries can be found) to a new feature space so that a linear decision boundary can be constructed to separate the data in the transformed space. However, this feature transformation approach raises a few issues about high feature dimensionality and high computational requirements. Alternatively, nonlinear classifiers can be created by applying a procedure similar to the linear ones to construct maximum margin hyperplanes, except that every dot product in the transformed space is replaced by a kernel function in the original feature space. Computing the dot products using kernels is considerably cheaper than using the transformed features. Several different kernel functions have been proposed and, for text classification, it seems that the SVM with a simple linear kernel performs comparably to nonlinear alternatives (Joachims 1998). An SVM with a linear kernel is used in our evaluation.

3.2.4 Augmented latent semantic indexing spaces

Latent semantic indexing (LSI) (Deerwester et al. 1990) is a well-known information retrieval technique. By deploying a rank-reduced feature–document space through the singular value decomposition (SVD) (Golub and van Loan 1996), it effectively transforms individual documents into their semantic content vectors to estimate the major associative patterns of features and documents and to diminish the obscuring noise in feature usage (Berry et al. 1995).

LSI can be used as a learning algorithm for spam filtering by replacing the notion of query relevance with the notion of category membership. An experiment of this approach on the Ling-Spam corpus was reported in Gee (2003) and it constructs a single LSI space to accommodate both spam and legitimate email training data. This simple application has some drawbacks (Jiang 2006). LSI itself is a completely unsupervised learning algorithm and when it is applied to (supervised) spam filtering, valuable category discriminative information embedded in training data should be extracted and integrated in model learning to boost classification accuracy. There are several approaches that can be used toward this goal. For instance, we can select distinctive features by exploring their category distributions (see Section 3.3) and introduce two separate LSI learning spaces (one for each email category). Feature selection also helps reduce computational requirements due to the SVD algorithm in the model.

For a given email training set, each of the two rank-reduced spaces can be constructed by using the data of its respective category and conceptually it would provide a more accurate category content profile than that produced from a single combined space. In practice, however, this dual-space approach may still encounter difficulties in classifying some email messages since many spam messages are purposely crafted to look legitimate and to mislead spam filters. This has been verified by our extensive experiments. In order to ameliorate this problem, a new model that uses augmented LSI learning spaces was proposed in Jiang (2006). More precisely, for each constructed category LSI space, this model augments the space with a small number of the training samples that are closest to the category in appearance but actually belong to the other category in label. This augmented LSI space model can effectively help classify those difficult target messages correctly, which are similar to the augmented samples used in the training, while maintaining accurate classification of other messages.

Expansion of the augmented training samples is carried out by cluster centroids. For each email category, we construct one or multiple clusters. For each cluster c_j, its centroid is computed as

$$a_{c_j} = \frac{1}{k}\sum_{i=1}^{k} d_{n_i}, \ d_{n_i} \in c_j, \tag{3.11}$$

and it can be used to represent the most important topic covered in the cluster (Jiang 2006). Once the cluster centroids of a category c are identified, all training samples from the other category are compared against the centroids and the most similar ones are then chosen to add to the training set of c. Selecting the sizes of clusters and augmented samples of a category can vary depending on the data to be learned. The cluster size can also be set by a silhouette plot (Kaufman and Rousseeuw 1990) on a given training dataset. In our experiments, we use the augmented sample sizes of 18 and 70 for the corpora PU1 and ZH1 (see Section 3.5), respectively.

To use two separate augmented LSI spaces for classification, several approaches have been considered and evaluated in Jiang (2006) that coordinate and classify target email messages into their respective classes. For a given target message, the first approach simply projects it onto both LSI spaces and then uses the most semantically similar training sample to decide the class for the message. The second approach classifies the message similarly but by applying a fixed number of the top most similar training samples in the spaces and using either the sum or average of computed similarity values from both classes to make its classification decision. The third approach is a hybrid one that intends to combine the ideas of the first two methods and also to mollify some of their shortcomings. Essentially, it determines the class for the target message by linearly balancing the votes or decisions made by the first two methods. Experiments indicate that in general the hybrid approach delivers significantly better classification results (Jiang 2006) and it is used in the study.

3.2.5   Radial basis function networks

RBF networks have many applications in science and engineering and can also be used to build learning models for filtering spam email (Jiang 2007). A typical RBF network has a feedforward connected structure of three layers: an input layer, a hidden layer of nonlinear processing neurons, and an output layer (Bishop 1995). For email classification, the input layer of the network has n neurons and it takes input training samples d. The hidden layer contains k computational neurons; each neuron can be mathematically described by an RBF ϕ_i that maps a distance between two vectors in the Euclidean norm into a real value:

$$\phi_i(x) = \phi(\|x - a_i\|_2), \ i = 1, 2, \ldots, k, \tag{3.12}$$

where a_i are the RBF centers in the input sample space and, in general, k is less than the size of training samples. The output layer of the network has two neurons that produces the target message category according to

$$c_j = \sum_{i=1}^{k} w_{ij}\phi_i(x), \quad j = 1, 2, \tag{3.13}$$

where w_{ij} is the weight connecting the ith neuron in the hidden layer to the jth neuron in the output layer. The neuron activation ϕ_i is a nonlinear function of the distance; the closer the distance, the stronger the activation. The most commonly used basis function is the Gaussian

$$\phi(x) = e^{-\frac{x^2}{2\sigma^2}}, \tag{3.14}$$

where σ is a width parameter that controls smoothness properties of the basis function.

In the spam filtering model (Jiang 2007), the network parameters, i.e. centers, widths, and weights, are set by a two-stage training procedure, which is computationally efficient. The first stage of training is to form a representation of the density distribution in input space in terms of the parameters of the RBFs. The centers a_i and widths σ are determined by relatively fast and unsupervised clustering algorithms, clustering each email category independently to obtain k basis functions for the category. In general, the larger the value of k, the better the classification outcomes and, of course, the higher the cost it carries in network training. With the computed and fixed centers and widths for the hidden layer, the second stage of training selects the weights of the output layer by a logistic regression procedure. Once all network parameters are determined, the model can be deployed to target email messages for classification, and classification outcomes from the network are computed by a weighted sum of the hidden layer activations, as is shown in Equation (3.13).

Recently, an RBF-based semi-supervised text classifier has also been developed (Jiang 2009). It integrates a clustering-based expectation maximization algorithm into the RBF training process and can learn for classification from a very small number of labeled training samples and a large pool of additional unlabeled data effectively.

3.3 Data preprocessing

In this section, we begin with some data preprocessing procedures that include feature selection and message representation, followed by a discussion of classification effectiveness measures for spam filtering.

3.3.1 Feature selection

As in general text classification, appropriate feature selection can be quite useful in aiding email classification. A term or feature is referred to as a word, a number, or a symbol in an email message. In spam filtering, features from training samples are selected according to their contributions to profiling legitimate or spam messages and those unselected features are removed from the data for model learning and deployment. The objectives of feature selection are twofold. On one hand, it is designed for dimensionality reduction in the message feature space. Dimensionality reduction aims to trim down the number of features to be modeled while the content of individual messages is still preserved. It generally helps speed up a model training process. On the other hand, feature selection intends to filter out irrelevant features, helping build an accurate and effective model for spam filtering. This is particularly valuable to certain machine-learning algorithms such as RBF networks, which treat every data feature equally in their distance computations and therefore are somewhat incapable of distinguishing relevant features from irrelevant ones.

Two steps of feature selection are used in our experiments. First, for a given set of training data, features are extracted and selected with an unsupervised setting. This is carried out by removing the stop or common words and applying a word stemming procedure. Then, the features with low message frequencies or low corpus frequencies are eliminated from the training data, as these features may not help much in differentiating messages for categories and may add some obscuring noise in email classification. The selection process also removes those features with very high corpus frequencies in the training data as many of these features distribute almost equally between spam and legitimate categories and may not be valuable in characterizing the email categories. Next, features are selected by their frequency distributions between spam and legitimated training messages. This supervised feature selection procedure intends, using those labeled training samples, to further identify the features that distribute most differently between the categories.

There are several supervised feature selection methods that have been widely used in text classification (Sebastiani 2002). They include the chi-square statistic (CHI), information gain (IG), and odds ratio (OR) criteria. The IG criterion quantifies the amount of information gained for category prediction by knowledge of the presence or absence of a feature in a message. More precisely, IG of a feature t about a category c can be expressed as

$$IG(t, c) = \sum_{c' \in \{c, \bar{c}\}} \sum_{t' \in \{t, \bar{t}\}} P(t', c') \log \frac{P(t', c')}{P(t')P(c')}, \qquad (3.15)$$

where $P(c')$ and $P(t')$ denote the probability that a message belongs to category c' and the probability that a feature t' occurs in a message, respectively, and $P(t', c')$ is the joint probability of t' and c'. All probabilities can be estimated by frequency counts from the training data. Another popular feature selection method is CHI. It measures the lack of independence between the occurrence of feature t and the occurrence of class c. In other words, features are ranked with respect to the quantity

$$CHI(t, c) = \frac{n[P(t, c)P(\bar{t}, \bar{c}) - P(t, \bar{c})P(\bar{t}, c)]^2}{P(t)P(\bar{t})P(c)(\bar{c})}, \qquad (3.16)$$

where n is the size of training data D (see Section 3.2) and the probability notations have the same interpretations as in Equation (3.15). For instance, $P(\bar{c})$ represents the probability that a message does not belong to category c. The third feature selection criterion, OR, has also been used in text classification and it measures the ratio of the odds of term t occurring in a message of class c to the odds of the term not occurring in c and can be defined as

$$OR(t, c) = \frac{P(t|c)(1 - P(t|\bar{c}))}{(1 - P(t|c))P(t|\bar{c})}. \qquad (3.17)$$

The effectiveness of the feature selection methods for text classification has been studied and compared, e.g. by Yang and Pedersen (1997), and some experiments with the criteria described above have also been conducted in this study. Among these three feature selection methods, our experiments suggest that the IG measure produces more stable classification results, so we used it in the selection process.

Through feature selection, the feature dimensionality of a training dataset can be reduced significantly. For instance, in the experiments with PU1 (see Section 3.5.1) the original feature size of the corpus, which is over 20 000 can be trimmed down to tens, hundreds, and thousands.

3.3.2   Message representation

After feature selection, each message is encoded as a numeric vector whose elements are the values of the retained feature set. Each feature value is associated with a local and global feature weight, representing the relative importance of the feature in the message and the overall importance of the feature in the corpus, respectively. Our experiments indicate that feature frequencies are more informative than a simple binary coding (which, for instance, is used in Zhang et al. (2004)) in the context of email classification.

There are several choices to weight a feature or term locally and globally based on its frequencies. For a given term t and document d, the traditional 'log(tf)–idf' term weight is defined as

$$w_{t,d} = \log(1 + tf_{t,d}) \log \frac{|D|}{df_t}, \qquad (3.18)$$

where $tf_{t,d}$ is the term frequency (tf) of t in d, df_t is the document frequency (df) of t, or the number of documents in a collection D that contain t, and $|D|$ is the size of the collection. The second component on the right hand side of Equation (3.18) represents the inverse document frequency (idf) of t. This term weighting scheme is used in this work and it produces good classification results.

3.4   Evaluation of email classification

The effectiveness of a text classifier can be evaluated in terms of its precision (p) and recall (r) measures. For a classifier and with respect to a category c, if the numbers of true positive, false positive, and false negative decisions on category c from the classifier are tp, fp, and fn, respectively, then the precision and recall are defined as

$$p = \frac{tp}{tp + fp}, \quad r = \frac{tp}{tp + fn}. \tag{3.19}$$

In brief, the precision measure is gauged by the percentage of documents classified to c which actually are, whereas the recall is quantified by the percentage of documents from c that are categorized by the classifier. Clearly, these two quantities trade off against one another and one single measure that balances both is the F-measure, which is the weighted harmonic mean of precision and recall. With an equal weight for both precision and recall, we have the commonly used F_1 measure

$$F_1 = \frac{2pr}{p + r}. \tag{3.20}$$

All these effectiveness measures, however, do not take a possible unbalanced misclassification cost into consideration. Spam email filtering can be a cost-sensitive learning process in the sense that misclassifying a legitimate message to spam (*false positive*) is typically a more severe error than misclassifying a spam message to legitimate (*false negative*). In reality, if a legitimate message is mistakenly classified and placed into a user's trash-mail box, then the user may not find this out for a short or long period of time and, depending on how important the message is, a delayed reading of the message could come with some negative consequences. In our experiments, an accuracy measure that uses a weight λ to reflect the unbalanced cost between false positive and false negative errors, or the weighted accuracy (Androutsopoulos et al. 2004), is used as the effectiveness criterion and it can be defined as

$$WA(\lambda) = \frac{\lambda tn + tp}{\lambda(tn + fp) + (tp + fn)}, \tag{3.21}$$

where the quantities tp, fp, and fn are the same as in Equation (3.19), tn denotes the true negative classification count, and λ is a cost parameter. The WA formula assumes that a false positive error is λ times more costly than a false negative one. We use $\lambda = 1$ for the case where both false positive and false negative errors have an equal cost and also a value of λ that is greater than one, such as nine, to indicate a higher cost of false positive errors. It is still arguable if such a higher cost in spam filtering can be quantified by a simple constant (Hidalgo 2002), and the cost should perhaps depend on several variable external factors. In this study, we use $\lambda = 9$ (or any other number in a similar quantity) just as a value to illustrate whether or not and how the effectiveness of the algorithms may change when a cost-sensitive condition is imposed.

3.5   Experiments

In this section, we use two benchmark email testing corpora to compare the efficacy of the five machine-learning algorithms, discussed in Section 3.2, for spam email filtering and provide the experimental results and analysis. Note that the input data to the classifiers is the preprocessed message vectors after both feature selection and feature weighting.

3.5.1   Experiments with PU1

PU1 is a benchmark spam testing corpus that contains a total of 1099 real email messages received by a single email user over a certain period of time (Androutsopoulos et al. 2004) and it is partitioned into 618 legitimate and 481 spam messages. The messages in the corpus have been preprocessed with all attachments, HTML tags, and header fields, except for subject lines which were removed, and the retained words in the email subject line and body text were encoded numerically for privacy protection.

There are a few other publicly accessible spam datasets such as the 2005 TREC spam corpus that can be used for spam filtering evaluation. However, most of them were aggregated from multiple different email sources or recipients, and some of the large ones were constructed by simply adding some newly gathered email messages to what had been collected. For very understandable privacy reasons, it has been a challenge for IT researchers to find coherent, reliable, and updated public email data, which can reflect what an average email user receives, for conducting experiments and producing meaningful and comparable testing results.

It should be pointed out that, in this study, we use only email subject line and body text as the email content. This is a constraint imposed by construction of the corpora we used in the experiments. The machine-learning algorithms investigated in this chapter, however, can plainly be applied to broader email content. As noted by several previous studies, e.g. Zhang et al. (2004), the features from other email text such as headers are indeed useful in discriminating spam email. Therefore, we expect that the classification accuracy of the algorithms presented in this section would be further increased if we were to use the broader content that includes email header fields.

The experiments on PU1 are performed using 10-fold cross-validation. That is, the corpus is partitioned into 10 equally sized subsets and each experiment takes one subset for testing and the remaining ones for training and the process repeats 10 times with each subset taking a turn for testing. The effectiveness is then evaluated by averaging over the 10 experiments, delivered as an average weighted accuracy defined in Equation (3.21). Various feature sizes are also used in the experiments that range from 50 to 1650 with an increment of 100.

Classification effectiveness of the five algorithms, measured by the average weighted accuracy over all feature sizes that have been considered, is shown in Figure 3.1 ($\lambda = 1$) and Figure 3.2 ($\lambda = 9$), respectively. The case of $\lambda = 1$ may reflect classification efficacy of the algorithms for general cost-insensitive learning with a small number of classes. Figure 3.1 shows that RBF performs very well over small feature sizes, but, along with LB, it produces less accurate classification than all other three classifiers at large feature sizes. On the other hand, LSI behaves in a fairly opposite way: it is the least accurate classifier over small feature sizes but achieves good accuracy at large feature sizes. The relatively stable performance of NB, SVM, and LB through all feature sets can be observed, where NB is the top performer, followed closely by SVM and then LB at a distance.

Now, we turn to the case of $\lambda = 9$ and we intend to use the generated weighted accuracy values to demonstrate whether or not and how the accuracy results of an algorithm change when a false positive error is to be punished more than a false negative error or a cost-sensitive condition is imposed. The changes, if any, should ultimately depend on how well the algorithm can profile legitimate messages and make small numbers of false positive errors. For both NB and LB classifiers, their accuracy values in this case are not significantly different from those in Figure 3.1 and, relatively, their false positive errors are comparable to their false negative ones. Similar observations can also be made for SVM. On the

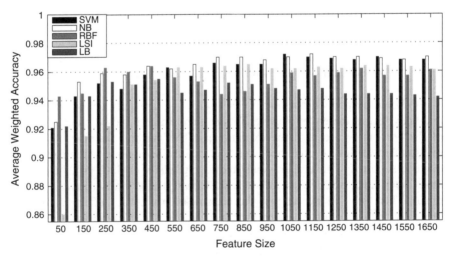

Figure 3.1   Average weighted classification accuracy with $\lambda = 1$ (PU1).

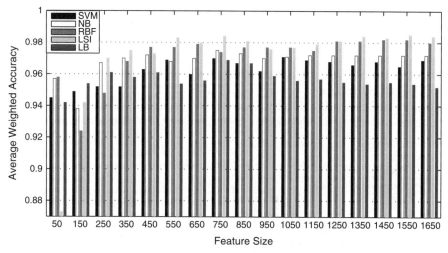

Figure 3.2   Average weighted classification accuracy with λ = 9 *(PU1).*

other hand, since LSI, followed very closely by RBF, carries somewhat smaller numbers of false positive errors than other classifiers, its accuracy values are lifted for it to become the top performer. A detailed analysis of LSI and RBF on their error counts suggests that a richer feature set generally helps the classifiers characterize legitimate messages and improve classification of the category. But it may not be useful for them to improve their classification of spam messages. One possible explanation for this phenomenon may be related to the vocabularies used in the respective email categories. It is hypothesized that spam email has a strong correspondence between a small set of features and the category, while legitimate email likely carries more sophisticated characteristics. The spam category could attain good classification with a small vocabulary while the legitimate category requires a large vocabulary, which can be assisted by feature expansion.

3.5.2   Experiments with ZH1

In this subsection, we present the experiments of the five classifiers on a Chinese spam corpus ZH1 (Zhang et al. 2004). The experiments aim to demonstrate the capability of individual classifiers to classifying email written in a language with a different linguistic structure. Chinese text does not have explicit word boundaries like English, and words in the text can be extracted by some specially designed word segmentation software (Zhang et al. 2004). The construction of corpus ZH1 is very similar to PU1 where ZH1 is made up of 1205 spam and 428 legitimate email messages. All messages in the corpus are also numerically encoded. Note that, in contrast to PU1, ZH1 has more spam email than legitimate email in the corpus and this helps examine whether or not and how the classifiers are possibly influenced in their model learning by unbalanced training sample sizes between the categories. Experiments on ZH1 are also performed using 10-fold cross-validation and the same feature sets as those with PU1.

Figure 3.3 and Figure 3.4 show the average weighted accuracy values obtained by all five classifiers over the feature sizes for $\lambda = 1$ and $\lambda = 9$, respectively. For the case of equal misclassification cost ($\lambda = 1$), Figure 3.3 indicates that SVM and LB perform best over most feature sizes, followed by LSI and then RBF; in this case, NB evidently fails to be comparable. When a higher cost on false positive errors is considered ($\lambda = 9$), similar observations can be made from

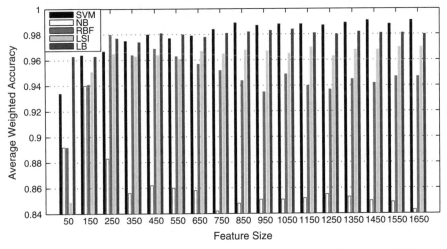

Figure 3.3   Average weighted classification accuracy with $\lambda = 1$ (ZH1).

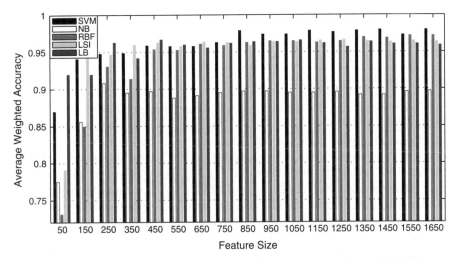

Figure 3.4   Average weighted classification accuracy with $\lambda = 9$ (ZH1).

Figure 3.4, but this time, at those feature sizes that are greater than 350, both LSI and RBF become much more competitive than LB and SVM. All four of these classifiers achieve high classification accuracy.

3.6 Characteristics of classifiers

In comparison to general text classification, spam email filtering represents a special, cost-sensitive, and very challenging classification task. It has two categories to be classified. The cost of the two types of misclassification errors is different and many spam messages are purposely and carefully constructed to look very much like legitimate ones. Though both spam and legitimate email messages may have a similar appearance, there may be still some important and different characteristics for each email category that should not be overlooked. For instance, in contrast to spam email, legitimate email has in general a broader vocabulary and also perhaps more eclectic subject matter. Ideally, a successful machine-learning algorithm used in this particular classification domain should fully utilize potential differences between the email categories and, more importantly, should be capable of profiling legitimate messages accurately and carry only a small number of false positive misclassification errors.

As in many other applications of machine learning, declaring one algorithm as the best for spam filtering is a difficult task and perhaps almost impossible. The experiments and analysis conducted in this study, however, have revealed some interesting characteristics among the five classifiers investigated. They are summarized below.

Naive Bayes (NB). This classifier is simple and the fastest in model learning among the five classifiers. It can work well for text classification. Since the algorithm assumes that individual features are completely independent of one another, the classifier can benefit from effective feature selection, which is demonstrated in the PU1 experiments. In the same vein, NB can perform poorly if it is applied to a dataset where there are some observable dependencies among features. One possible explanation for the inadequate performance of NB on ZH1 is the language on which the corpus is based. Chinese is a language with a vast vocabulary and it is extremely difficult to automatically extract meaningful words or features correctly from a Chinese document; many Chinese words are also polysemous (the words can have very different meanings depending on the context in which they are used). All of these language characteristics may contribute to inaccurate probability estimation and heavy feature dependencies, which can inevitably reduce the power of the NB algorithm.

LogitBoost (LB). As a boosting algorithm, LB combines multiple simple base learners (decision stumps in this case) iteratively to make a powerful classifier. Although the base learner has a very simple structure, the ensemble construction can still be very time consuming. The success of LB on text classification or spam filtering seems to depend on the dataset but generally LB delivers competitive results. One interesting and unique characteristic of the method is its insensibility to feature size and large feature sizes may not help improve its classification accuracy. Hence, it seems that a relatively small feature size such as 250 could be used for the model training. Finally, the learning ability of the classifier for profiling a category appears to be influenced by the size of available training samples of the category.

SVM. As reported by several previous studies, SVM is a very stable classifier and is also scalable to feature dimensionality. In this study, SVM consistently performs as the best or as a very competitive classifier, in particular when cost-insensitive classification is considered. The linear SVM used in this study is also relatively fast in model training.

Augmented latent semantic indexing spaces (LSI). The LSI model constructs two separate rank-reduced and augmented learning spaces, one for each email category. In this study, the model has been demonstrated to be a very reliable classifier and it consistently delivers competitive classification results. The model also seems well suited to cost-sensitive spam filtering and this could be in part due to its integrated clustering component for constructing the augmented LSI spaces. Good performance of the classifier generally requires a feature size of about 500 or larger. Algorithm training can be expensive if the feature size becomes very large.

Radial basis function networks. The RBF-based classifier performs reasonably well, especially when it is evaluated as a cost-sensitive learning algorithm. This is likely contributed by the clustering process used in its first stage of network training. The model's performance appears to be affected by the clustering accuracy and, in addition, the classifier seems to be sensitive to feature size, so any excessive feature selection attempts should be avoided.

Overall, in terms of adaptability to cost-sensitive spam filtering, the classifiers based on LSI and RBF demonstrate their strength in this evaluation. Although these are two quite different machine-learning algorithms, they share one common characteristic: that is, both use a clustering component in their model training. Since clustering can potentially group messages by topics, an integrable clustering process can benefit from machine-learning algorithms in enhancing their profile accuracy of legitimate email (i.e. the category with a large vocabulary), and in reducing their numbers of false positive errors.

3.7   Concluding remarks

In this chapter, we provide an evaluation study of five current machine-learning algorithms proposed for spam filtering. The algorithms are described and compared by using various feature sizes, determined through an effective feature selection procedure, and by conducting experiments on some benchmark spam testing corpora constructed from two different languages. In particular, this study evaluates the adaptability of the algorithms for cost-sensitive spam filtering and, in this regard, the classifiers based on augmented LSI spaces, SVM, and RBF networks are the top performers. The experimental results also suggest that the newly proposed LSI and RBF classifiers represent two very competitive alternatives to other well-known methods for text and spam classification.

Content-based spam email filtering is a challenging classification task and success of the process can practically be influenced by many choices that include the selection of the algorithm, data and data preprocessing, feature selection, and decision criteria. In this study, we use only the email subject line and body text as the content for learning. For future work, we plan to expand the email content for spam filtering by the features contained in header fields, which seem to be reliable and useful (Zhang et al. 2004). Also, we plan to revisit some machine-learning algorithms to further improve their classification effectiveness on cost-sensitive learning. For instance, we would like to see how an optimal number of clusters for the LSI and RBF classifiers can be determined to create an accurate representation of topics among messages of both email categories.

3.8   Acknowledgements

This work was in part supported by a faculty research grant from the University of San Diego.

References

Aha W and Albert M 1991 Instance-based learning algorithms. *Machine Learning* **6**, 37–66.

Androutsopoulos I, Paliouras G and Michelakis E 2004 Learning to filter unsolicited commercial e-mail. Technical Report, NCSR Demokritos.

Berry M, Dumais S and O'Brien W 1995 Using linear algebra for intelligent information retrieval. *SIAM Review* **37**(4), 573–595.

Bishop C 1995 *Neural Networks for Pattern Recognition*. Oxford University Press.

Christianini B and Shawe-Taylor J 2000 *An Introduction to Support Vector Machines and Other Kernel-based Learning Methods*. Cambridge University Press.

Deerwester S, Dumais S, Furnas G, Landauer T and Harshman R 1990 Indexing by latent semantic analysis. *Journal of the American Society for Information Science* **41**, 391–409.

Drucker H, Wu D and Vapnik V 1999 Support vector machines for spam categorization. *IEEE Transactions on Neural Networks* **10**, 1048–1054.

Friedman J, Hastie T and Tibshirani R 2000 Additive logistic regression: A statistical view of boosting. *Annals of Statistics* **28**(2), 337–374.

Gee K 2003 Using latent semantic indexing to filter spam. *Proceedings of the ACM Symosium on Applied Computing*, pp. 460–464.

Golub G and van Loan C 1996 *Matrix Computations*, third edn. Johns Hopkins University Press.

Hidalgo J 2002 Evaluating cost-sensitive unsolicited bulk email categorization. *Proceedings of the 17th ACM Symposium on Applied Computing*, pp. 615–620.

Jiang E 2006 Learning to semantically classify email messages. *Lecture Notes in Control and Information Sciences* **344**, 700–711.

Jiang E 2007 Detecting spam email by radial basis function networks. *International Journal of Knowledge-based and Intelligent Engineering Systems* **11**, 409–418.

Jiang E 2009 Semi-supervised text classification using RBF networks. *Lecture Notes in Computer Science* **5772**, 95–106.

Joachims T 1998 Text categorization with support vector machines – learning with many relevance features. *Proceedings of the 10th European Conference on Machine Learning*, pp. 137–142.

Kaufman L and Rousseeuw P 1990 *Finding Groups in Data*. John Wiley & Sons, Inc.

Manning C, Raghavan P and Schutze H 2008 *Introduction to Information Retrieval*. Cambridge University Press.

Mitchell T 1997 *Machine Learning*. McGraw-Hill.

Platt J 1999 Fast training of support vector machines using sequential minimal optimization. In *Advances in Kernel Methods: Support Vector Learning* (ed. Scholkop B, Burges C and Smola A) MIT Press pp. 185–208.

Rocchio J 1997 Relevance feedback information retrieval In *The Smart Retrieval System: Experiments in automatic document processing* (ed. Salton G) Prentice Hall pp. 313–323.

Sahami M, Dumais S, Heckerman D and Horvitz E 1998 A Bayesian approach to filtering junk e-mail. *Proceedings of AAAI Workshop*, pp. 55–62.

Sebastiani F 2002 Machine learning in automated text categorization. *ACM Computing Surveys* **1**, 1–47.

Witten T and Frank E 2005 *Data Mining*, second edn. Morgan Kaufmann.

Yang Y and Pedersen J 1997 A comparative study on feature selection in text categorization. *Proceedings of the 14th International Conference on Machine Learning*, pp. 412–420.

Zhang H 2004 The optimality of naive bayes. *Proceedings of the 17th International FLAIRS Conference*.

Zhang L, Zhu J and Yao T 2004 An evaluation of statistical spam filtering techniques. *ACM Transactions on Asian Language Information Processing* **3**, 243–369.

4

Utilizing nonnegative matrix factorization for email classification problems

Andreas G. K. Janecek and Wilfried N. Gansterer

4.1 Introduction

About a decade ago, unsolicited bulk email ('spam') started to become one of the biggest problems on the Internet. A vast number of strategies and techniques were developed and employed to fight email spam, but none of them can be considered a final solution to this problem. In recent years, phishing ('password fishing') has become a severe problem in addition to spam email. The term covers various criminal activities which try to fraudulently acquire sensitive data or financial account credentials from Internet users, such as account user names, passwords, or credit card details. Phishing attacks use both social engineering and technical means. In contrast to unsolicited but harmless spam email, phishing is an enormous threat for all big Internet-based commercial operations.

Generally, email classification methods can be categorized into three groups, according to their point of action in the email transfer process. These groups are pre-send methods, post-send methods, and new protocols, which are based on modifying the transfer process itself. Pre-send methods, which act before the email is transported over the network, are very important because of their

Text Mining: Applications and Theory   edited by Michael W. Berry and Jacob Kogan
© 2010, John Wiley & Sons, Ltd

potential to avoid the wasting of resources caused by spam. However, since the efficiency of these methods depends on their widespread deployment, most of the currently used email filtering techniques belong to the group of post-send methods. Amongst others, this group comprises techniques such as black-, white-, and graylisting, or rule-based filters, which block email based on a predetermined set of rules. Using these rules, features describing an email message can be extracted. After extracting the features, a classification process can be applied to predict the class (ham, spam, phishing) of unclassified email. An important approach for increasing the speed of the classification process is to perform feature subset selection (removal of redundant and irrelevant features) or dimensionality reduction (use of low-rank approximations of the original data) prior to the classification.

Low-rank approximations replace a large and often sparse data matrix with a related matrix of much lower rank. The objective of these techniques – which can be utilized in many data mining applications such as image processing, drug discovery, or text mining – is to reduce the required storage space and/or to achieve more efficient representations of the relationship between data elements. Depending on the approximation technique used, great care must be taken in terms of storage requirements. If the original data matrix is very sparse (as is the case for many text mining problems), the storage requirements for the reduced rank matrices might be higher than for the original data matrix with higher dimensions (since the reduced rank matrices are often almost completely dense). Besides well-known techniques like principal component analysis (PCA) and singular value decomposition (SVD), there are several other low-rank approximation methods like vector quantization (Linde et al. 1980), factor analysis (Gorsuch 1983), QR decomposition (Golub and Van Loan 1996) or CUR decomposition (Drineas et al. 2004). In recent years, another approximation technique for *nonnegative* data has been used successfully in various fields. The *nonnegative matrix factorization* (NMF, see Section 4.2) determines reduced rank nonnegative factors $\mathbf{W}$ and $\mathbf{H}$ which approximate a given nonnegative data matrix $\mathbf{A}$, such that $\mathbf{A} \approx \mathbf{WH}$.

In this chapter, we investigate the application of NMF to the task of email classification. We consider the interpretability of the NMF factors in the email classification context and try to take advantage of information provided by the basis vectors in $\mathbf{W}$ (interpreted as basis emails or the basis features). Motivated by this context, we also investigate a new initialization technique for NMF based on ranking the original features. This approach is compared to standard random initialization and other initialization techniques for NMF described in the literature. Our approach shows faster reduction of the approximation error than random initialization and comparable results to existing but often more time-consuming approaches. Moreover, we analyze classification methods based on NMF. In particular, we introduce a new method that combines NMF with LSI (Latent Semantic Indexing) and compare this approach to standard LSI.

4.1.1   Related work

The utilization of low-rank approximations in the context of email classification has been analyzed in Gansterer et al. (2008b). In this work, LSI was applied successfully both on purely textual features and on features extracted by rule-based filtering systems. Especially the features from rule-based filters allowed for a strong reduction of the dimensionality without losing significant accuracy in the classification process. Feature reduction is particularly important if time constraints play a role, as in the online processing of email streams. In Gansterer et al. (2008a) a framework for such situations was presented – an enhanced self-learning variant of graylisting (temporarily rejecting email messages) was combined with a reputation-based trust mechanism to separate SMTP communication from feature extraction and classification. This architecture minimizes the workload on the client side and achieves very high spam classification rates. A comparison of the classification accuracy achieved with feature subset selection and low-rank approximation based on PCA in the context of email classification can be found in Janecek et al. (2008).

Nonnegative matrix factorization. Paatero and Tapper (1994) published an article on *positive matrix factorization*, but the work by Lee and Seung (1999) five years later achieved much more popularity and is known as a standard reference for NMF. The two NMF algorithms introduced in Lee and Seung (1999) – *multiplicative update algorithm* and *alternating least squares* (Berry et al. 2007; Lee and Seung 2001) – provide good baselines against which newer algorithms (e.g. the *gradient descent* algorithm) have to be judged.

NMF initialization. All algorithms for computing the NMF are iterative and require initialization of **W** and **H**. While the general goal – to establish initialization techniques and algorithms that lead to better overall error at convergence – is still an open issue, some initialization strategies can improve the NMF in terms of faster convergence and faster error reduction. Although the benefits of good NMF initialization techniques are well known in the literature, rather few algorithms for non-random initializations have been published so far.

Wild et al. (Wild 2002; Wild et al. 2003, 2004) were among the first to investigate the initialization problem of NMF. They used spherical k-means clustering based on the centroid decomposition (Dhillon and Modha 2001) to obtain a structured initialization for **W**. More precisely, they partition the columns of **A** into k clusters and select the centroid vectors for each cluster to initialize the corresponding columns in **W**. Their results show faster error reduction than random initialization, thus saving expensive NMF iterations. However, since this decomposition must run a clustering algorithm on the columns of **A**, it is expensive as a preprocessing step (cf. Langville et al. (2006)).

Langville et al. (2006) also provided some new initialization ideas and compared the aforementioned centroid clustering approach and random seeding to four new initialization techniques. While two algorithms (Random Acol and Random C) only slightly decrease the number of NMF iterations and another algorithm (Co-occurrence) turns out to contain very expensive computations, the *SVD–Centroid* algorithm clearly reduces the approximation error and therefore the number of NMF iterations compared to random initialization. The algorithm initializes $\mathbf{W}$ based on a SVD–centroid decomposition (Wild 2002) of the low-dimensional SVD factor $\mathbf{V}_{n \times k}$, which is much faster than a centroid decomposition on $\mathbf{A}_{m \times n}$ since $\mathbf{V}$ is much smaller than $\mathbf{A}$. Nevertheless, the SVD factor $\mathbf{V}$ must be available for this algorithm, and the computation of $\mathbf{V}$ can obviously be time consuming.

Boutsidis and Gallopoulos (2008) initialized $\mathbf{W}$ and $\mathbf{H}$ using a technique called *nonnegative double singular value decomposition* (NNDSVD) which is based on two SVD processes, one approximating the data matrix $\mathbf{A}$ (rank k approximation) and the other approximating positive sections of the resulting partial SVD factors. The authors performed various numerical experiments and showed that NNDSVD initialization is better than random initialization in terms of faster convergence and error reduction in all test cases, and generally appears to be better than the centroid initialization in Wild (2002).

4.1.2   Synopsis

This chapter is organized as follows. In Section 4.2 we review some basics of NMF and make some comments on the interpretability of the basis vectors in $\mathbf{W}$ in the context of email classification ('basis features' and 'basis emails'). We also provide some information about the data and feature sets used in this chapter. Some ideas about new NMF initialization techniques are discussed in Section 4.3, and Section 4.4 focuses on new classification methods based on NMF. We conclude our work in Section 4.5.

4.2   Background

In this section, we review the definition and characteristics of NMF and give a brief overview of the two NMF algorithms considered in this work, as well as their termination criteria and computational complexity. We then describe the datasets used for experimental evaluation and make some remarks on the interpretability of the NMF factors $\mathbf{W}$ and $\mathbf{H}$ in the context of email classification problems.

4.2.1   Nonnegative matrix factorization

NMF (Lee and Seung 1999; Paatero and Tapper 1994) consists of reduced rank nonnegative factors $\mathbf{W} \in \mathbb{R}^{m \times k}$ and $\mathbf{H} \in \mathbb{R}^{k \times n}$ with (problem-dependent) $k \ll \min\{m, n\}$ that approximate a given nonnegative data matrix $\mathbf{A} \in \mathbb{R}^{m \times n}$ so that $\mathbf{A} \approx \mathbf{WH}$. Despite the fact that the product $\mathbf{WH}$ is only an approximate factorization of $\mathbf{A}$ of rank at most k, $\mathbf{WH}$ is called a nonnegative matrix factorization of **A**. The nonlinear optimization problem underlying NMF can generally be stated as

$$\min_{W,H} f(\mathbf{W}, \mathbf{H}) = \frac{1}{2}||\mathbf{A} - \mathbf{WH}||_F^2, \qquad (4.1)$$

where $||.||_F$ is the Frobenius norm. Although the Frobenius norm is commonly used to measure the error between the original data **A** and **WH**, other measures are also possible, e.g. an extension of the Kullback–Leibler divergence to positive matrices (Dhillon and Sra 2006). Unlike the SVD, the NMF is not unique, and convergence is not guaranteed for all NMF algorithms. If they converge, then they usually converge to local minima only (potentially different ones for different algorithms). Fortunately, the data compression achieved with only local minima has been shown to be of desirable quality for many data mining applications (Langville et al. 2006).

Due to its nonnegativity constraints, NMF produces so-called 'additive parts-based' (or 'sum-of-parts') representations of the data (in contrast to many other linear representations such as SVD, PCA, or ICA (Independent Component Analysis)). This is an impressive benefit of NMF, since it makes the interpretation of the NMF factors much easier than for factors containing positive and negative entries, and enables a non-subtractive combination of parts to form a whole (Lee and Seung 1999). For example, the features in **W** (called 'basis vectors') may be topics of clusters in textual data, or parts of faces in image data. Another favorable consequence of the nonnegativity constraints is that both factors **W** and **H** are often naturally sparse (see, e.g., the update steps of the alternating least squares algorithm below, where negative elements are set to zero).

4.2.2   Algorithms for computing NMF

NMF algorithms can be divided into three general classes: multiplicative update (MU), alternating least squares (ALS), and gradient descent (GD) algorithms. A review of these three classes can be found in Berry et al. (2007). In this chapter, we use implementations of the MU and ALS algorithms (these algorithms do not depend on a step size parameter, as is the case for GD) from the Statistics Toolbox v6.2 in MATLAB (included since the R2008a release). The termination criteria for both algorithms were also adapted from the MATLAB implementation. Pseudo code for the general structure of NMF algorithms is given in Algorithm 1.

Algorithm 1 – General structure of NMF algorithms

1: given matrix $\mathbf{A} \in \mathbb{R}^{m \times n}$ with $k \ll \min\{m, n\}$:
2: **for** *rep* = 1 to *maxrepetition* **do**
3:    **W** = rand(m, k);
4:    **H** = rand(k, n);
5:    **for** i = 1 to *maxiter* **do**

6:        perform NMF update steps
7:        check termination criterion
8:    **end for**
9: **end for**

Most algorithms need pre-initialized factors $\mathbf{W}$ and $\mathbf{H}$, but some algorithms (e.g. the ALS algorithm) only need one pre-initialized factor. The standard ALS algorithm uses a pre-initialized $\mathbf{W}$, but the algorithm also works with a pre-initialized factor $\mathbf{H}$ (in this case, lines 1 and 3 in Algorithm 3 have to be exchanged). In the basic form of most NMF algorithms, the factors are initialized randomly. Different update steps are briefly described in the following.

Multiplicative update. The update steps for the MU algorithm given in Lee and Seung (2001) are based on the mean squared error objective function. Adding ε in each iteration avoids division by zero. A typical value used in practice is $\varepsilon = 10^{-9}$.

Algorithm 2 – Update steps for the MU algorithm

1: $\mathbf{H} = \mathbf{H} . * (\mathbf{W}^T \mathbf{A}) ./(\mathbf{W}^T \mathbf{W} \mathbf{H} + \varepsilon)$;
2: $\mathbf{W} = \mathbf{W} . * (\mathbf{A} \mathbf{H}^T) ./(\mathbf{W} \mathbf{H} \mathbf{H}^T + \varepsilon)$;

Alternating least squares algorithm. ALS algorithms were first mentioned in Paatero and Tapper (1994). In an alternating manner, a least squares step is followed by another least squares step. In this rather simple case, all negative elements resulting from the least squares computation are set to 0 to ensure nonnegativity. The standard ALS algorithm only needs to initialize the factor $\mathbf{W}$; the factor $\mathbf{H}$ is computed in the first iteration.

Algorithm 3 – Update steps for the ALS Algorithm

1: solve for $\mathbf{H} : \mathbf{W}^T \mathbf{W} \mathbf{H} = \mathbf{W}^T \mathbf{A}$;
2: set all negative elements in $\mathbf{H}$ to 0;
3: solve for $\mathbf{W} : \mathbf{H} \mathbf{H}^T \mathbf{W}^T = \mathbf{H} \mathbf{A}^T$;
4: set all negative elements in $\mathbf{W}$ to 0;

Both algorithms are iterative and depend on the initialization of $\mathbf{W}$ (and $\mathbf{H}$). Since the iterates generally converge to a local minimum, often several instances of the algorithm are run using different random initializations, and the best of the solutions is chosen. A proper nonrandom initialization of $\mathbf{W}$ and/or $\mathbf{H}$ (depending on the algorithm) can avoid the need to repeat several factorizations. Moreover, it may speed up convergence of a single factorization and reduce the error as defined in Equation (4.1).

Termination criterion

Generally, the termination criterion for NMF algorithms comprises three components. The first condition is based on the maximum number of iterations (the algorithm iterates until the maximal number of iterations is reached). The second condition is based on the required approximation accuracy (if the approximation error in Equation (4.1) drops below a predefined threshold, the algorithm stops). Finally, the third condition is based on the relative change of the factors **W** and **H** from one iteration to another. If this change is below a predefined threshold δ, then the algorithm also terminates.

Computational complexity of NMF

A single update step of the MU algorithms has the complexity $\mathcal{O}(kmn)$ (since **A** is $m \times n$, **W** is $m \times k$, and **H** is $k \times n$), see, for example, Li et al. (2007) and Robila and Maciak (2009). Considering the number of iterations i of the NMF yields an overall complexity of $\mathcal{O}(ikmn)$. For the ALS algorithm, the complexity for solving the equations in lines 1 and 3 of Algorithm 3 need to be considered additionally. In its most general form, these equations are solved using an orthogonal triangular factorization.

4.2.3   Datasets

The datasets used for evaluation consist of 15 000 email messages, divided into three groups – ham, spam, and phishing. The email messages were taken partly from the *Phishery*[1] and partly from the 2007 TREC corpus.[2] The email messages are described by 133 features. A part of these features is purely text based, other features comprise online features and features extracted by rule-based filters. Some of the features specifically test for spam messages, while other features specifically test for phishing messages. As a preprocessing step we scaled all feature values to [0,1] to ensure that they have the same range.

The structure of phishing messages tends to differ significantly from the structure of spam messages, but it may be quite close to the structure of regular ham messages (because for a phishing message it is particularly important to look like a regular message from a trustworthy source). A detailed discussion and evaluation of this feature set has been given in Gansterer and Pölz (2009).

The email corpus was split into two sets (for training and for testing), the training set consisting of the oldest 4000 email messages of each class (12 000 messages in total), and the test set consisting of the newest 1000 email messages of each class (3000 messages in total). This chronological ordering of historical data allows for simulation of the changes and adaptations in spam and phishing messages which occur in practice. Both email sets are ordered by the classes – the first group in each set consists of ham messages, followed by spam and phishing

---

[1] http://phishery.internetdefence.net
[2] http://trec.nist.gov/data/spam.html

messages. Due to the nature of the features, the data matrices are rather sparse. The larger (training) set has 84.7% zero entries, and the smaller (test) set has 85.5% zero entries.

4.2.4   Interpretation

A key characteristic of NMF is the representation of basis vectors in **W** and the representation of basis coefficients in the second NMF factor **H**. With these coefficients the columns of **A** can be represented in the basis given by the columns of **W**. In the context of email classification, **W** may contain *basis features* or *basis emails*, depending on the structure of the original data. If NMF is applied to an *email × feature* matrix (i.e. every row in **A** corresponds to an email message), then **W** contains k basis *features*. If NMF is applied on the transposed matrix (*feature × email* matrix, i.e. every column in **A** corresponds to an email message), then **W** contains k basis *email messages*.

Basis features. Figure 4.1 shows three basis features $\in \mathbb{R}^{12\,000}$ (for $k = 3$) for our training set when NMF is applied to an *email × feature* matrix. The three different groups of objects – ham (first 4000 messages), spam (middle 4000 messages), and phishing (last 4000 messages) – are easy to identify. The group of phishing emails tends to yield high values for basis feature 1, while basis feature 2 shows the highest values for the spam messages. The values of basis feature 3 are generally smaller than those of basis features 1 and 2, and this basis feature is clearly dominated by the ham messages.

Basis email messages. The three basis email messages $\in \mathbb{R}^{133}$ (again for $k = 3$) resulting from NMF on the transposed (*feature × email*) matrix are plotted

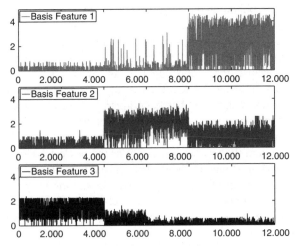

Figure 4.1   Basis features for k = 3.

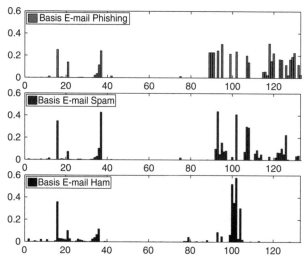

Figure 4.2   Basis email messages for k = 3.

in Figure 4.2. The figure shows two features (16 and 102) that have a relatively high value in all basis emails, indicating that these features do not distinguish well between the three classes of email. Other features better distinguish between classes. For example, features 89–91 and 128–130 have a high value in basis email 1, and are (close to) zero in the other two basis emails. Investigation of the original data shows that these features tend to have high values for phishing email, indicating that the first basis email represents a phishing message. Using the same procedure, the third basis email can be identified to represent ham messages (indicated by features 100 and 101). Finally, basis email 2 represents spam.

This rich structure observed in the basis vectors should be exploited in the context of classification methods. However, the structure of the basis vectors heavily depends on the concrete feature set used. In the following, we discuss the application of feature selection techniques in the context of NMF initialization.

4.3   NMF initialization based on feature ranking

As already mentioned in Section 4.1.1, the initialization of the NMF factors has a big influence on the speed of convergence and the error reduction of NMF algorithms. Although the benefits of good initialization are well known, randomized seeding of **W** and **H** is still the standard approach for many NMF algorithms. Existing approaches, such as initialization based on spherical k-means clustering (Wild 2002) or nonnegative double singular value decomposition (NNDSVD) (Boutsidis and Gallopoulos 2008) can be rather time consuming. Obviously, the trade-off between the computational cost in the initialization step and the computational cost of the actual NMF algorithm needs to be balanced carefully. In some situations, an expensive preprocessing step may overwhelm the cost savings in the subsequent NMF update steps. In the following, we introduce a simple and fast initialization step based on feature subset selection and show comparisons with random initialization and the NNDSVD approach mentioned earlier.

4.3.1 Feature subset selection

The main idea of feature subset selection (FS) is to rank features according to how well they differentiate between object classes. Redundant or irrelevant features can be removed from the dataset as they can lead to a reduction of classification accuracy or clustering quality and to an unnecessary increase of computational cost. The output of the FS process is a ranking of features based on the applied FS algorithm. The two FS methods used in this chapter are based on *information gain* and *gain ratio*, both reviewed briefly in the following.

Information gain. One option for ranking the features of email messages according to how well they differentiate the three classes ham, spam, and phishing is to use their *information gain*, which is also used to compute splitting criteria for decision trees. The overall entropy I of a given dataset S is defined as

$$I(S) := -\sum_{i=1}^{C} p_i \log_2 p_i, \tag{4.2}$$

where C denotes the total number of classes and p_i the portion of instances that belong to class i. The reduction in entropy or the *information gain* is computed for each attribute A according to

$$IG(S, A) := I(S) - \sum_{v \in A} \frac{|S_{A,v}|}{|S|} I(S_{A,v}), \tag{4.3}$$

where v is a value of A and $S_{A,v}$ is the set of instances where A has value v.

Gain ratio. Information gain favors features which assume many *different* values. Since this property of a feature is not necessarily connected with the splitting information of a feature, we also ranked the features based on their *information gain ratio*, which normalizes the information gain and is defined as $GR(S, A) := IG(S, A)/splitinfo(S, A)$, where

$$splitinfo(S, A) := -\sum \frac{|S_{A,v}|}{|S|} \log_2 \frac{|S_{A,v}|}{|S|}. \tag{4.4}$$

4.3.2 FS initialization

After determining the feature ranking based on information gain and gain ratio, we use the k first ranked features to initialize $\mathbf{W}$ (denoted as *FS initialization* in the following). Since feature selection aims at reducing the feature space, our initialization is applied in the setup where **W** contains basis features (i.e. every row in **A** corresponds to an email message, cf. Section 4.2.4). FS methods are usually computationally inexpensive (see, e.g., Janecek et al. (2008) for a comparison of information gain and PCA runtimes) and can thus be used as a computationally cheap but effective initialization step. A detailed runtime comparison of information gain, gain ratio, NNDSVD, random seeding, and other initialization methods as well as the initialization of **H** (at the moment **H** is randomly seeded) are work in progress.

Results. Figures 4.3 and 4.4 show the NMF approximation error for our new initialization strategy for both information gain (infogain) and gain ratio feature ranking as well as for NNDSVD and random initialization when using the ALS algorithm. As a baseline, the figures also show the approximation error based on an SVD of **A**, which gives the best possible rank k approximation of **A**. For rank $k = 1$, all NMF variants achieve the same approximation error as the SVD, but for higher values of k the SVD has a smaller approximation error than the NMF variants (as expected, since SVD gives the best rank k approximation in terms of approximation error). Note that when the maximum number of iterations inside a single NMF factorization (*maxiter*) is high (*maxiter* = 30 in Figure 4.4), the approximation errors are very similar for all initialization strategies used and are very close to the best approximation computed with SVD. On the other hand, with a small number of iterations (*maxiter* = 5 in Figure 4.3), it is clearly visible that random seeding cannot compete with initialization based on NNDSVD and feature selection. Moreover, for this small value of *maxiter*, the FS initializations (both information gain and gain ratio ranking) show better error

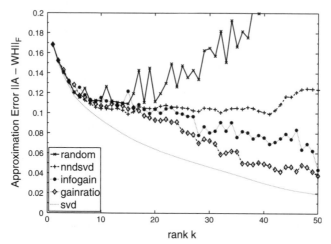

Figure 4.3   Approximation error for different initialization strategies and varying rank k *using the ALS algorithm (* maxiter = 5*).*

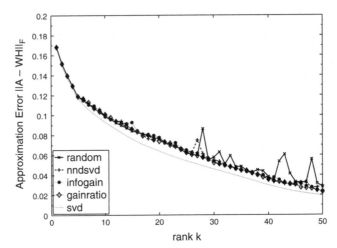

Figure 4.4   Approximation error for different initialization strategies and varying rank k *using the ALS algorithm (* maxiter = *30).*

reduction than NNDSVD with increasing rank k. For higher values of *maxiter* the gap between the different initialization strategies decreases until the error curves become basically identical when *maxiter* is about 30 (see Figure 4.4).

Runtime. In this subsection we analyze runtimes for computing NMF for different values of rank k and different values of *maxiter* using the ALS algorithm. All runtime comparisons in this chapter were measured on a SUN Fire X4600M2 with eight AMD quad-core Opteron 8356 processors (32 cores overall) with 2.3 GHz CPU and 64 GB of memory. Since the MATLAB implementation of the ALS algorithm is not the best implementation in terms of runtime, we computed the ALS update steps (see Algorithm 3) using an economy-size QR factorization: that is, only the first n columns of the QR factorization factors **Q** and **R** are computed (here n is the smaller dimension of the original data matrix **A**). This saves computation time (about 3.7 times faster than the original ALS algorithm implemented in MATLAB), but achieves identical results to the MATLAB implementation. The algorithms terminated when the number of iterations exceeded the predefined threshold *maxiter*; that is, the approximation error was not integrated in the stopping criterion. Consequently, the runtimes do not depend on the initialization strategy used (neglecting the marginal runtime savings due to sparse initializations). In this setup, a linear relationship between runtime and the rank of k can be observed. Reducing the number of iterations (lower values of *maxiter*) brings important reductions in runtimes. This underlines the benefits of our new initialization techniques. As Figure 4.3 has shown, our FS initialization reduces the number of iterations required for achieving a certain approximation error compared to existing approaches.

Table 4.1 compares runtimes needed to achieve different approximation error thresholds with different values of *maxiter* for different initialization strategies.

Table 4.1 Runtime comparison.

$\|A-WH\|_F$	maxiter 5	maxiter 10	maxiter 15	maxiter 20	maxiter 25	maxiter 30
			Gain ratio initialization			
0.10	0.6 s ($k = 17$)	1.0 s ($k = 11$)	1.5 s ($k = 11$)	2.0 s ($k = 11$)	2.2 s ($k = 10$)	2.7 s ($k = 10$)
0.08	0.9 s ($k = 27$)	1.5 s ($k = 22$)	2.2 s ($k = 21$)	2.9 s ($k = 19$)	3.1 s ($k = 19$)	3.3 s ($k = 19$)
0.06	1.1 s ($k = 32$)	2.0 s ($k = 30$)	2.8 s ($k = 28$)	3.7 s ($k = 28$)	4.6 s ($k = 27$)	5.4 s ($k = 26$)
0.04	1.5 s ($k = 49$)	2.4 s ($k = 40$)	3.9 s ($k = 40$)	5.0 s ($k = 40$)	6.3 s ($k = 38$)	7.2 s ($k = 38$)
			Information gain initialization			
0.10	0.6 s ($k = 18$)	1.0 s ($k = 12$)	1.6 s ($k = 12$)	1.8 s ($k = 10$)	2.2 s ($k = 10$)	2.7 s ($k = 10$)
0.08	1.0 s ($k = 28$)	1.5 s ($k = 22$)	2.3 s ($k = 22$)	2.9 s ($k = 19$)	3.1 s ($k = 19$)	3.3 s ($k = 19$)
0.06	1.5 s ($k = 48$)	2.0 s ($k = 30$)	3.0 s ($k = 30$)	3.7 s ($k = 28$)	4.6 s ($k = 28$)	5.4 s ($k = 26$)
0.04	1.6 s ($k = 50$)	2.5 s ($k = 42$)	4.1 s ($k = 42$)	5.1 s ($k = 41$)	6.3 s ($k = 38$)	7.2 s ($k = 38$)
			NNDSVD initialization			
0.10	0.6 s ($k = 15$)	1.0 s ($k = 12$)	1.6 s ($k = 12$)	1.8 s ($k = 10$)	2.2 s ($k = 10$)	2.7 s ($k = 10$)
0.08	n.a.	1.7 s ($k = 25$)	2.6 s ($k = 25$)	2.6 s ($k = 18$)	3.1 s ($k = 19$)	3.2 s ($k = 18$)
0.06	n.a.	2.1 s ($k = 32$)	3.1 s ($k = 32$)	3.9 s ($k = 29$)	4.6 s ($k = 28$)	5.7 s ($k = 30$)
0.04	n.a.	n.a.	n.a.	5.1 s ($k = 41$)	6.3 s ($k = 38$)	7.2 s ($k = 38$)
			Random initialization			
0.10	n.a.	0.9 s ($k = 10$)	1.4 s ($k = 10$)	1.8 s ($k = 10$)	2.2 s ($k = 10$)	2.7 s ($k = 10$)
0.08	n.a.	1.5 s ($k = 22$)	2.3 s ($k = 22$)	2.5 s ($k = 17$)	3.1 s ($k = 19$)	3.3 s ($k = 19$)
0.06	n.a.	n.a.	n.a.	4.1 s ($k = 31$)	4.5 s ($k = 26$)	5.4 s ($k = 26$)
0.04	n.a.	n.a.	n.a.	5.4 s ($k = 45$)	6.7 s ($k = 42$)	7.3 s ($k = 39$)

Obviously, a given approximation error $||\mathbf{A} - \mathbf{WH}||_F$ can be achieved much faster with small *maxiter* and high rank k than with high *maxiter* and small rank k. As can be seen in Table 4.1, an approximation error of 0.04 or smaller can be computed in 1.5 and 1.6 seconds, respectively, when using gain ratio and information gain initialization (here, only five iterations (*maxiter*) are needed to achieve an approximation error of 0.04). To achieve the same approximation error with NNDSVD or random initialization, more than 5 seconds are needed (here, 20 iterations are needed to achieve the same approximation error).

4.4   NMF-based classification methods

In this section we investigate new classification algorithms which utilize NMF for developing a classification model. First, we look at the classification accuracy achieved with the basis features in $\mathbf{W}$ when initialized with the techniques explained in Section 4.3. Since, in this case, NMF is computed on the complete data, this technique can only be applied on data that is already available before the classification model is built.

In the second part of this section we introduce a classifier based on NMF which can be applied dynamically to new email data. We present a combination of NMF with LSI and compare it to standard LSI based on SVD.

4.4.1   Classification using basis features

Figures 4.5 and 4.6 show the overall classification accuracy for a ternary classification problem (ham, spam, phishing) using different values of *maxiter* for all four initialization strategies mentioned in Section 4.3. As the classification algorithm we used a support vector machine (SVM) with a radial basis kernel provided by the MATLAB LIBSVM (v2.88) interface (Chang and Lin 2001). For the results shown in this section, we performed fivefold cross-validation on the larger email corpus (consisting of 12 000 email messages, cf. Section 4.2.3).

The results based on the four NMF initialization techniques (infogain, gainratio, nndsvd, and random) were achieved by applying an SVM on the rows of $\mathbf{W}$, where every email message is described by k basis *features*, i.e. every column of $\mathbf{W}$ corresponds to a basis feature (cf. Section 4.2.4). As NMF algorithm we used multiplicative update (MU). For comparison to the original features, we applied a standard SVM classification on the email messages characterized by k best ranked information gain features (SVMinfogain). The graph for SVMinfogain is identical in both figures since the *maxiter* factor in the NMF algorithm has no influence on the result.

Classification results. For lower ranks ($k < 30$), the SVMinfogain results are markedly below the results achieved with nonrandomly initialized NMF (infogain, gainratio, and nndsvd). This is not very surprising, since $\mathbf{W}$ contains compressed information about all features (even for small ranks of k). Random NMF

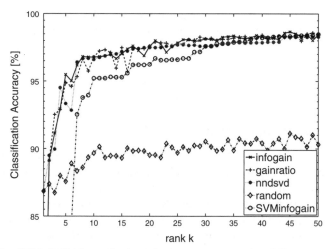

Figure 4.5   SVM (RBF kernel) classification accuracy for different initialization methods using the MU algorithm (maxiter = 5*).*

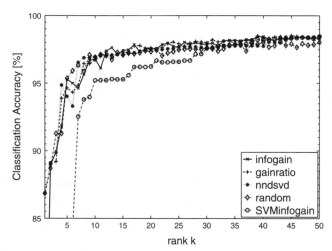

Figure 4.6   SVM (RBF kernel) classification accuracy for different initialization methods using the MU algorithm (maxiter = 30*).*

initialization of **W** (random) achieves even lower classification accuracy for *maxiter* = 5 (see Figure 4.5). The classification result remains unsatisfactory even for large values of k. With larger *maxiter* (cf. Figure 4.6), the classification accuracy for randomly seeded **W** increases and achieves results comparable to infogain, gainratio, and nndsvd. Comparing the results of the FS initialization and nndsvd initialization, it can be seen that there is no large gap in the classification accuracy. We would like to point out the clear decline in the classification accuracy of nndsvd for $k = 6$ (in both figures). Surprisingly, the classification results for *maxiter* = 5 are only slightly worse than for *maxiter* = 30, which is in contrast to the approximation error results shown in Section 4.3. Consequently, fast (and accurate) classification is possible for small *maxiter* and small k (e.g. the average classification accuracy over infogain, gainratio, and nndsvd is 96.75% for $k = 10$ and *maxiter* = 5, compared to 98.34% for $k = 50$, *maxiter* = 50).

4.4.2   Generalizing LSI based on NMF

Now we look at the classification process in a dynamic setting where newly arriving email messages are to be classified. Obviously, this is not suitable for computing a new NMF for every new incoming email message. Instead, a classifier is constructed by applying NMF on a training sample and using the information provided by the factors $\mathbf{W}$ and $\mathbf{H}$ in the classification model. In the following, we present adaptations of LSI based on NMF and compare them to standard LSI (based on SVD). Note that in this section our datasets are transposed compared to the experiments in Sections 4.3 and 4.4.1. Hence, every column of $\mathbf{A}$ corresponds to an email message.

Review of VSM and standard LSI. A VSM (Raghavan and Wong 1999) is a widely used algebraic model where objects and queries are represented as vectors in a potentially very high-dimensional metric vector space. Generally speaking, given a query vector $\mathbf{q}$, the distances of $\mathbf{q}$ to all objects in a given *feature × object* matrix $\mathbf{A}$ can be measured (for example) in terms of the cosines of the angles between $\mathbf{q}$ and the columns of $\mathbf{A}$. The cosine φ_i of the angle between $\mathbf{q}$ and the ith column of $\mathbf{A}$ can be computed as

$$\text{(VSM)} : \cos\varphi_i = \frac{e_i^\top A^\top q}{||Ae_i||_2||q||_2}. \tag{4.5}$$

LSI (Langville 2005) is a variant of the basic VSM. Instead of the original matrix $\mathbf{A}$, the SVD is used to construct a low-rank approximation $\mathbf{A}_k$ of $\mathbf{A}$, such that $\mathbf{A} = \mathbf{U}\Sigma\mathbf{V}^\top \approx \mathbf{U}_k\Sigma_k\mathbf{V}_k^\top =: \mathbf{A}_k$. When $\mathbf{A}$ is replaced with $\mathbf{A}_k$, then the cosine of φ_i for the angle between $\mathbf{q}$ and the ith column of $\mathbf{A}$ is approximated as

$$\text{(SVD-LSI)} : \cos\varphi_i \approx \frac{e_i^\top V_k\Sigma_k U_k^\top q}{||U_k\Sigma_k V_k^\top e_i||_2||q||_2}. \tag{4.6}$$

Since some terms on the right hand side of this equation only need to be computed once for different queries ($\mathbf{e}_i^\top \mathbf{V}_k\Sigma_k$ and $||\mathbf{U}_k\Sigma_k\mathbf{V}_k^\top \mathbf{e}_i||_2$), LSI saves storage and computational cost. Further, the approximated data often gives a cleaner and more efficient representation of the relationship between data elements (Langville et al. 2006) and can uncover *latent* information in the data.

NMF-based classifiers. We investigate two novel concepts for using NMF as a low-rank approximation within LSI (see Figure 4.7). The first approach, which

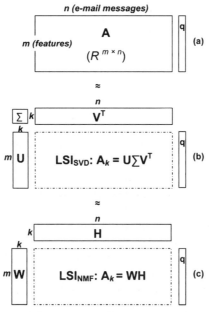

Figure 4.7   Overview: (a) basic VSM; (b) LSI using SVD; (c) LSI using NMF.

we call NMF-LSI, simply replaces the approximation within LSI with a different approximation. Instead of using $U_k \Sigma_k V_k^T$, we approximate A with $A_k := W_k H_k$ from the rank k NMF. Note that when using NMF, the value of k must be fixed prior to the computation of W and H. The cosine of the angle between q and the ith column of A can then be approximated as

$$\text{(NMF-LSI)} : \cos \varphi_i \approx \frac{e_i^{\top} H_k^{\top} W_k^{\top} q}{||W_k H_k e_i||_2 ||q||_2}. \qquad (4.7)$$

To save computational cost, the leftmost term in the denominator and the leftmost part of the numerator (both involving W_k and H_k) can be computed a priori.

The second classifier. which we call NMF-BCC (NMF Basis Coefficient Classifier), is based on the idea that the basis coefficients in H can be used to classify new email. These coefficients are representations of the columns of A in the basis given by W. If W, H, and q are given, we can calculate a column vector x that minimizes the equation

$$\min_x ||Wx - q||. \qquad (4.8)$$

Since x is the best representation of q in the basis given by W, we search for the column of H which is closest to x for assigning q to one of the three classes of email. Moreover, the residual in Equation (4.8) indicates how close q is to the email messages in $\mathbf{A}$. The cosine of the angle between $\mathbf{q}$ and the ith column of $\mathbf{H}$ can be approximated as

$$(\text{NMF-BCC}) : \cos\varphi_i \approx \frac{e_i^\top H^\top x}{||He_i||_2 ||x||_2}. \tag{4.9}$$

It is obvious that the computation of the cosines in Equation (4.9) is much faster than for both other LSI variants mentioned earlier (since usually $\mathbf{H}$ is a much smaller matrix than $\mathbf{A}$), but the computation of $\mathbf{x}$ causes additional cost. These aspects will be discussed further at the end of this section.

Classification results. A comparison of the results achieved with LSI based on SVD (SVD-LSI), LSI based on NMF (NMF-LSI), the basis coefficient classifier (NMF-BCC), and a basic VSM (VSM) is shown in Figures 4.8 and 4.9, again for different values of *maxiter*. In contrast to Section 4.4.1, where we performed a cross-validation on the larger email corpus, here we used the big corpus as the training set and tested with the smaller corpus consisting of the 1000 *newest* email messages of each class. For classification, we considered the column of $\mathbf{A}$ with the smallest angle to $\mathbf{q}$ (no majority count) to assign $\mathbf{q}$ to one of the classes ham, spam, and phishing. The results shown in this section were achieved with random initialization.

Obviously, there is a big difference in the classification accuracy achieved with the NMF approaches for small and larger values of *maxiter*. With *maxiter* = 5 (see Figure 4.8), the NMF variants can hardly compete with LSI based on SVD and VSM. However, when *maxiter* is increased to 30, all NMF variants except

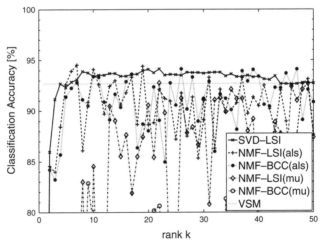

Figure 4.8 Classification accuracy for different LSI variants and VSM (maxiter = 5).

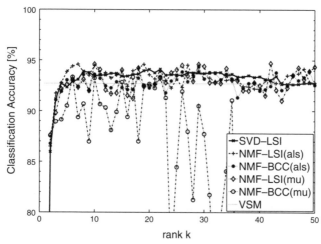

Figure 4.9 Classification accuracy for different LSI variants and VSM (maxiter = *30).*

NMF-BCC(mu) show comparable results (see Figure 4.9). For many values of k, the NMF variants achieved better classification accuracy than a basic VSM with all original features. Moreover, the standard ALS variant (NMF-LSI(als)) achieves very comparable results to LSI based on SVD, especially for small values of rank k (between 5 and 10). Note that this improvement of a few percent is substantial in the context of email classification. Moreover, as discussed in Section 4.2.4, the purely nonnegative linear representation within NMF makes the interpretation of the NMF factors much easier than that for the standard LSI factors. It is interesting to note that initialization of the factors **W** and **H** does not improve the classification accuracy when using the NMF-LSI and NMF-BCC classifiers. This is in contrast to the previous sections – especially when *maxiter* is small, the initialization was important for the SVM.

Runtimes. The computational runtime for all LSI variants comprises two steps. Prior to the classification process, the low-rank approximations of SVD and NMF, respectively, have to be computed. Afterward, any newly arriving email message (a single query vector) has to be classified.

Figure 4.10 shows the runtimes needed for computing the low-rank approximations, and Figure 4.11 shows the runtimes for the classification process of a singly query vector. As already mentioned in Section 4.3.2, the NMF runtimes depend almost linearly on the value of *maxiter*. Figure 4.10 shows that for almost any a given rank k, the computation of an SVD takes much longer than an NMF factorization with *maxiter* = 5, but is faster than a factorization with *maxiter* = 30. For computing the SVD we used MATLAB's svds() function, which computes only the first k largest singular values and associated singular

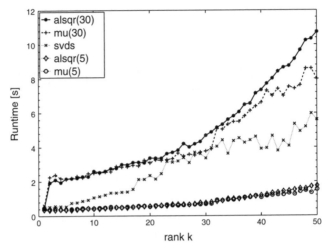

Figure 4.10   Runtimes for computing low-rank approximations based on SVD and variants of NMF of a 12 000 × 133 matrix (alsqr(30) *refers to the ALS algorithm computed with explicit QR factorization and* maxiter *set to 30).*

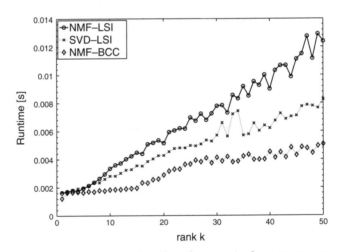

Figure 4.11   Runtimes for classifying a single query vector.

vectors of a matrix. The computation of the complete SVD usually takes much longer (but is not needed in this context). There is only a small difference in the runtimes for computing the ALS algorithm (using the economy-size QR factorization, cf. Section 4.3.2) and the MU algorithm, and, of course, no difference between the NMF-LSI and the NMF-BCC runtimes (since the NMF factorization has to be computed identically for both approaches). The difference in the computational cost between NMF-LSI and NMF-BCC is embedded in the classification process of query vectors, not in the factorization process of the training data.

Looking at the classification runtimes in Figure 4.11, it can be seen that the classification process using the basis coefficients (NMF-BCC) is faster than for SVD-LSI and NMF-LSI. Although the classification times for a single email are modest, they have to be considered for every single email that is classified. The classification (performed in MATLAB) of all 3000 email messages in our test sample took about 36 seconds for NMF-LSI, 24 seconds for SVD-LSI, and only 13 seconds for NMF-BCC (for rank $k = 50$).

Rectangular versus square data. Since the dimensions of the email data matrix used in this work are very imbalanced ($12\,000 \times 133$), we also compared runtime and approximation errors for data of the same size, but with balanced dimensions. We created square random matrices of dimension $\sqrt{133 \times 12\,000} \approx 1263$ and performed experiments on them identical to those in the previous section.

Figure 4.12 shows the runtime needed to compute the first k largest singular values and associated singular vectors for SVD (again using the svds() function from MATLAB) as well as the two NMF factorizations with different values of *maxiter*. For square **A**, the computation of the SVD takes much longer than for unbalanced dimensions. In contrast, both NMF approximations can be computed much faster (cf. Figure 4.10). For example, the computation of an SVD of rank $k = 50$ takes about eight times longer than the computation of an NMF of the same rank.

The approximation error for square random data is shown in Figure 4.13. The approximation error of both SVD and NMF is generally higher than for the email dataset (see Figures 4.3 and 4.4). It is interesting to note that the approximation error of the ALS algorithm decreases with increasing k until $k \approx 35$, and then increases again with higher values of k. Nevertheless, especially for smaller values of k, the ALS algorithm achieves an approximation error comparable to the SVD with much lower computational runtimes.

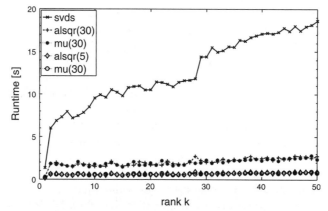

Figure 4.12 Runtimes for computing low-rank approximations based on SVD and variants of NMF of a random 1263 × 1263 matrix.

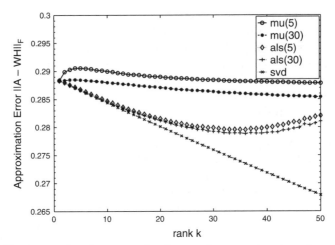

Figure 4.13   Approximation error for low-rank approximations based on SVD and variants of NMF on a random 1263 × 1263 matrix.

4.5   Conclusions

The application of nonnegative matrix factorization (NMF) to ternary email classification tasks (ham vs. spam vs. phishing messages) has been investigated. We have introduced a fast initialization technique based on feature subset selection (FS initialization) which significantly reduces the approximation error of the NMF compared to randomized seeding of the NMF factors **W** and **H**. Comparison of our approach to existing initialization strategies such as NNDSVD (Boutsidis and Gallopoulos 2008) shows basically the same accuracy when many NMF iterations are performed, and much better accuracy when the NMF algorithm is restricted to a small number of iterations.

Moreover, we investigated and evaluated two new classification methods which are based on NMF. We showed that using the basis features of **W** generally achieves much better results than using the original features. While the number of iterations (*maxiter*) in the iterative process for computing the NMF seems to be a crucial factor for the classification accuracy when random initialization is used, the classification results achieved with FS initialization and NNDSVD depend only weakly on this parameter, leading to high classification accuracy even for small values of *maxiter* (see Figures 4.5 and 4.6). This is in contrast to the approximation error illustrated in Figures 4.3 and 4.4, where the number of iterations is important for all initialization variants.

As a second classification method we constructed NMF-based classifiers to be applied on newly arriving email messages without recomputing the NMF. For this purpose, we introduced two LSI classifiers based on NMF (computed with the ALS algorithm) and compared them to standard LSI based on SVD. Both new variants achieved a classification accuracy comparable to standard LSI when using the ALS algorithm and can often be computed faster, especially when the dimensions of the original data matrix are close to each other (in this case, the computation of the SVD usually takes much longer than an NMF factorization).

A copy of the codes used in this chapter is available from the authors or at `http://rlcta.univie.ac.at`.

Future work. Our investigations indicate several important and interesting directions for future work. First of all, we will focus on analyzing the computational cost of various initialization strategies (FS initialization vs. NNDSVD etc.). Moreover, we will look at updating schemes for our NMF-based LSI approach, since for real-time email classification a dynamical adaptation of the training data (i.e. adding new email to the training set) is essential. We also plan to work on strategies for the initialization of **H** (currently, **H** is randomly initialized) for our FS initialization (Section 4.3) and the comparison of the MU and ALS algorithms to other NMF algorithms (gradient descent, algorithms with sparseness constraints, etc.).

4.6   Acknowledgements

We gratefully acknowledge financial support from the CPAMMS-Project (grant# FS397001) in the Research Focus Area 'Computational Science' of the University of Vienna. We also thank David Poelz for providing us with detailed information about the characteristics of the email features.

References

Berry MW, Browne M, Langville AN, Pauca PV and Plemmons RJ 2007 Algorithms and applications for approximate nonnegative matrix factorization. *Computational Statistics & Data Analysis* **52**(1), 155–173.

Boutsidis C and Gallopoulos E 2008 SVD based initialization: A head start for nonnegative matrix factorization. *Pattern Recognition* **41**(4), 1350–1362.

Chang CC and Lin CJ 2001 *LIBSVM: a library for support vector machines*. Software available at `http://www.csie.ntu.edu.tw/~cjlin/libsvm`.

Dhillon IS and Modha DS 2001 Concept decompositions for large sparse text data using clustering. *Machine Learning* **42**(1), 143–175.

Dhillon IS and Sra S 2006 Generalized nonnegative matrix approximations with bregman divergences. *Advances in Neural Information Processing Systems 18: Proceedings of the 2005 Conference*, pp. 283–290.

Drineas P, Kannan R and Mahoney MW 2004 Fast Monte Carlo algorithms for matrices III: Computing a compressed approximate matrix decomposition. *SIAM Journal on Computing* **36**(1), 184–206.

Gansterer WN and Pölz D 2009 E-mail classification for phishing defense. In *Advances in Information Retrieval, 31st European Conference on IR Research, ECIR 2009, Toulouse,*

France, April 6–9, 2009. Proceedings (ed. Boughanem M, Berrut C, Mothe J and Soulé-Dupuy C), vol. 5478 of *Lecture Notes in Computer Science*. Springer.

Gansterer WN, Janecek A and Kumer KA 2008a Multi-level reputation-based greylisting. *Proceedings of Third International Conference on Availability, Reliability and Security (ARES 2008)*, pp. 10–17. IEEE Computer Society, Barcelona, Spain.

Gansterer WN, Janecek A and Neumayer R 2008b Spam filtering based on latent semantic indexing. *In: Survey of Text Mining 2*, vol. 2, pp. 165–183. Springer.

Golub GH and Van Loan CF 1996 *Matrix Computations (Johns Hopkins Studies in Mathematical Sciences)*. The Johns Hopkins University Press.

Gorsuch RL 1983 *Factor Analysis* 2nd edn. Lawrence Erlbaum.

Janecek A, Gansterer WN, Demel M and Ecker GF 2008 On the relationship between feature selection and classification accuracy. *JMLR: Workshop and Conference Proceedings* **4**, 90–105.

Langville AN 2005 The linear algebra behind search engines. *Journal of Online Mathematics and its Applications (JOMA), 2005, Online Module*.

Langville AN, Meyer CD and Albright R 2006 Initializations for the nonnegative matrix factorization. *Proceedings of the 12th ACM SIGKDD International Conference on Knowledge Discovery and Data Mining*.

Lee DD and Seung HS 1999 Learning the parts of objects by non-negative matrix factorization. *Nature* **401**(6755), 788–791.

Lee DD and Seung HS 2001 Algorithms for non-negative matrix factorization. *Advances in Neural Information Processing Systems* **13**, 556–562.

Li X, Cheung WKW, Liu J and Wu Z 2007 A novel orthogonal NMF-based belief compression for POMDPs. *Proceedings of the 24th International Conference on Machine Learning*, pp. 537–544.

Linde Y, Buzo A and Gray RM 1980 An algorithm for vector quantizer design. *IEEE Transactions on Communications* **28**(1), 702–710.

Paatero P and Tapper U 1994 Positive matrix factorization: A non-negative factor model with optimal utilization of error estimates of data values. *Environmetrics* **5**(2), 111–126.

Raghavan VV and Wong SKM 1999 A critical analysis of vector space model for information retrieval. *Journal of the American Society for Information Science* **37**(5), 279–287.

Robila S and Maciak L 2009 Considerations on parallelizing nonnegative matrix factorization for hyperspectral data unmixing. *Geoscience and Remote Sensing Letters* **6**(1), 57–61.

Wild SM 2002 Seeding non-negative matrix factorization with the spherical k-means clustering. *Master's Thesis, University of Colorado*.

Wild SM, Curry JH and Dougherty A 2003 Motivating non-negative matrix factorizations. *Proceedings of the Eighth SIAM Conference on Applied Linear Algebra*.

Wild SM, Curry JH and Dougherty A 2004 Improving non-negative matrix factorizations through structured initialization. *Pattern Recognition* **37**(11), 2217–2232.

5

Constrained clustering with k-means type algorithms

Ziqiu Su, Jacob Kogan and Charles Nicholas

5.1 Introduction

Clustering is a fundamental data analysis task that has numerous applications in many disciplines. Clustering can be broadly defined as a process of partitioning a dataset into groups, or clusters, so that elements of the same cluster are more similar to each other than to elements of different clusters.

In many cases additional information about the desired type of clusters is available (e.g. Basu et al. (2009)). When incorporated into the clustering process this information may lead to better clustering results. Motivated by Basu et al. (2004) we consider *pairwise constrained clustering*. In pairwise constrained clustering, we may have information about pairs of vectors that may not belong to the same cluster (*cannot-links*), information about pairs of vectors that must belong to the same cluster (*must-links*), or both. (For the first introduction of constrained clustering with a focus on instance-level constraints see Wagstaff and Cardie (2000) and Wagstaff et al. (2001).)

We focus on three k-means type clustering algorithms and two different distance-like functions. The clustering algorithms are k-means (Duda et al. 2000), smoka (Teboulle and Kogan 2005), and spherical k-means (Dhillon and Modha 1999). The distance-like functions are 'reverse Bregman divergence' (see

Text Mining: Applications and Theory edited by Michael W. Berry and Jacob Kogan
© 2010, John Wiley & Sons, Ltd

e.g. Kogan (2007a)) and 'cosine' similarity (see e.g. Berry and Browne (1999)). We show that for these algorithms and distance-like functions the pairwise constrained clustering problem can be reduced to clustering with *cannot-link* constraints only. We substitute *cannot-link* constraints by penalty, and propose clustering algorithms that tackle clustering with penalties.

The chapter is organized as follows. In Section 5.2 we introduce basic notations, and briefly review batch and incremental versions of classical quadratic k-means. Section 5.3 presents the clustering algorithm equipped with Bregman divergences and constraints. We show by an example that a straightforward adoption of batch k-means may lead to erroneous results, and introduce a modification of incremental k-means that generates a sequence of partitions with improved quality. We show that *must-link* constraints can be eliminated (the elimination technique is based on the methodology proposed in Zhang et al. (1997)). When information about a large number of must-linked vectors is available, the proposed elimination technique may significantly reduce the size of the dataset. Section 5.4 introduces a smoka type clustering with constrains (see e.g. Teboulle and Kogan (2005) and Teboulle (2007)). Elimination of *must-link* constraints is based on results reported in Kogan (2007b). Section 5.5 presents spherical k-means with constraints. Numerical experiments that illustrate the usefulness of constraints are collected in Section 5.6. Brief conclusions and future research directions are given in Section 5.7.

5.2   Notations and classical k-means

The entries of a vector $\mathbf{a} \in \mathbf{R}^n$ are denoted by $(\mathbf{a}[1], \ldots, \mathbf{a}[n])^T$. The size of a finite set $\mathcal{A}$ is denoted by $|\mathcal{A}|$. For a set of m vectors $\mathcal{A} = \{\mathbf{a}_1, \ldots, \mathbf{a}_m\} \subset \mathbf{R}^n$, a prescribed subset $\mathcal{C}$ of $\mathbf{R}^n$, and a 'distance-like' function $d(\mathbf{x}, \mathbf{a})$ we define a centroid $\mathbf{c} = \mathbf{c}(\mathcal{A})$ of the set $\mathcal{A}$ as a solution of the minimization problem

$$\mathbf{c} = \arg\min \left\{ \sum_{\mathbf{a} \in \mathcal{A}} d(\mathbf{x}, \mathbf{a}), \ \mathbf{x} \in \mathcal{C} \right\}. \tag{5.1}$$

Examples of distance-like functions include the squared Euclidean distance $d(\mathbf{x}, \mathbf{a}) = \|\mathbf{x} - \mathbf{a}\|^2$, and the relative entropy (also known as Kullback–Leibler divergence) $d(\mathbf{x}, \mathbf{a}) = \sum_{i=1}^{n} \mathbf{a}[i] \log(\mathbf{a}[i]/\mathbf{x}[i])$. While in the case of $d(\mathbf{x}, \mathbf{a}) = \|\mathbf{x} - \mathbf{a}\|^2$ the set $\mathcal{C}$ may be the entire space, when $d(\mathbf{x}, \mathbf{a}) = \sum_{i=1}^{n} \mathbf{a}[i] \log(\mathbf{a}[i]/\mathbf{x}[i])$, the set $\mathcal{C}$ housing centroids $\mathbf{x}$ should be restricted to vectors with at least nonnegative entries (in many text mining applications $\mathbf{a}[i] \geq 0$).

The quality of the set $\mathcal{A}$ is denoted by $Q(\mathcal{A})$ and is defined by

$$Q(\mathcal{A}) = \sum_{i=1}^{m} d(\mathbf{c}, \mathbf{a}), \quad \text{where } \mathbf{c} = \mathbf{c}(\mathcal{A}) \tag{5.2}$$

(we set $Q(\emptyset) = 0$ for convenience). Let $\Pi = \{\pi_1, \ldots, \pi_k\}$ be a partition of $\mathcal{A}$, i.e.

$$\bigcup_i \pi_i = \mathcal{A}, \text{ and } \pi_i \cap \pi_j = \emptyset \text{ if } i \neq j.$$

We abuse notation and define the quality of the partition Π by

$$Q(\Pi) = Q(\pi_1) + \cdots + Q(\pi_k) = \sum_{i=1}^{k} \sum_{\mathbf{a} \in \pi_i} d(\mathbf{c}_i, \mathbf{a}), \text{ where } \mathbf{c}_i = \mathbf{c}(\pi_i). \quad (5.3)$$

We aim to find a partition $\Pi^{\min} = \{\pi_1^{\min}, \ldots, \pi_k^{\min}\}$ that *minimizes* the value of the objective function Q. The problem is known to be NP-hard (see e.g. Brucker (1978)) and we seek algorithms that generate 'reasonable' solutions.

It is easy to see that centroids and partitions are associated as follows:

1. Given a partition $\Pi = \{\pi_1, \ldots, \pi_k\}$ of the set $\mathcal{A}$ one can define the corresponding centroids $\{\mathbf{c}(\pi_1), \ldots, \mathbf{c}(\pi_k)\}$ by

$$\mathbf{c}(\pi_i) = \arg\min \left\{ \sum_{\mathbf{a} \in \pi_i} d(\mathbf{x}, \mathbf{a}), \ \mathbf{x} \in \mathcal{C} \right\}. \quad (5.4)$$

2. For a set of k 'centroids' $\{\mathbf{c}_1, \ldots, \mathbf{c}_k\}$ one can define a partition $\Pi = \{\pi_1, \ldots, \pi_k\}$ of the set $\mathcal{A}$ by

$$\pi_i = \{\mathbf{a} : \mathbf{a} \in \mathcal{A}, \ d(\mathbf{c}_i, \mathbf{a}) \leq d(\mathbf{c}_l, \mathbf{a}) \text{ for each } l = 1, \ldots, k\} \quad (5.5)$$

(we break ties arbitrarily). Note that, in general, $\mathbf{c}(\pi_i) \neq \mathbf{c}_i$.

The classical batch k-means algorithm is a procedure that iterates between the two steps described above to generate a partition Π' from a partition Π (Duda et al. 2000). While step 2 is straightforward, step 1 requires us to solve a constrained optimization problem. The degree of difficulty involved depends on the distance-like function $d(\cdot, \cdot)$ and the set $\mathcal{C}$. The entire procedure is essentially a gradient-based algorithm.

Incremental k-means is an iterative algorithm that seeks to change the cluster affiliation of one vector per iteration.

Definition 5.2.1 *A first variation of a partition Π is a partition Π' obtained from Π by removing a single vector $\mathbf{a}$ from a cluster π_i of Π and assigning this vector to an existing cluster π_j of Π.*

The decision of which vector to move is based on *exact* computation of the change in the objective. The change Δ in the objective Q caused by moving a vector $\mathbf{a}$ from cluster π_i to cluster π_j is given by

$$\Delta = \frac{|\pi_i|}{|\pi_i| - 1} \|\mathbf{c}(\pi_i) - \mathbf{a}\|^2 - \frac{|\pi_j|}{|\pi_j| + 1} \|\mathbf{c}(\pi_j) - \mathbf{a}\|^2 \quad (5.6)$$

(see e.g. (Kogan 2007a)).

Definition 5.2.2 *The partition* nextFV (Π) *is a first variation of* Π *so that for each first variation* Π' *one has*

$$Q\left(\text{nextFV}\left(\Pi\right)\right) \leq Q\left(\Pi'\right). \tag{5.7}$$

The computational complexity involved in finding the first variation does not exceed that required by the second step of batch k-means.

In the next section we show by an example that a straightforward application of batch k-means to clustering with cannot-link constraints may lead to erroneous results. The section suggests modifications of incremental k-means for constrained clustering of datasets equipped with Bregman distances.

5.3   Constrained k-means with Bregman divergences

We start with a detailed description of k-means constrained clustering and elimination of must-link constraints for a dataset equipped with squared Euclidean distance. At the end of the section the results are extended to Bregman distances.

5.3.1   Quadratic k-means with cannot-link constraints

We first focus on clustering with cannot-link constraints only. The constraints are substituted by a nonnegative penalty function, and a k-means like clustering is introduced on the dataset equipped with the penalty function. Partitioning of the constrained dataset and clustering of the dataset equipped with the penalty function are illustrated in Section 5.6.

For a vector set $\mathcal{A} = \{\mathbf{a}_1, \ldots, \mathbf{a}_m\} \subset \mathbf{R}^n$ and a symmetric penalty function $p :$ $\mathbf{R}^n \times \mathbf{R}^n \to \mathbf{R}_+$, $p(\mathbf{a}, \mathbf{a}) = 0$, $p(\mathbf{a}, \mathbf{a}') = p(\mathbf{a}', \mathbf{a})$, we define $Q(\mathcal{A})$, the quality of $\mathcal{A}$, as

$$Q(\mathcal{A}) = \sum_{\mathbf{a} \in \mathcal{A}} \|\mathbf{c} - \mathbf{a}\|^2 + \frac{1}{2} \sum_{\mathbf{a}, \mathbf{a}' \in \mathcal{A}} p(\mathbf{a}, \mathbf{a}'), \tag{5.8}$$

where $\mathbf{c}$ is the unique solution of

$$\min_{\mathbf{x}} \left\{ \sum_{\mathbf{a} \in \mathcal{A}} \|\mathbf{x} - \mathbf{a}\|^2 + \frac{1}{2} \sum_{\mathbf{a}, \mathbf{a}' \in \mathcal{A}} p(\mathbf{a}, \mathbf{a}') \right\},$$

which is given by the arithmetic mean of $\mathcal{A}$. Our aim is to identify an optimal k-cluster partition of $\mathcal{A}$.

Given a partition $\{\pi_1, \ldots, \pi_k\}$ and the corresponding centroids $\mathbf{c}_i$, it is tempting to adopt the two-stage batch k-means procedure with the following modification of Equation (5.5) that defines the new partition $\{\pi'_1, \ldots, \pi'_k\}$ as

$$\pi'_i = \left\{ \mathbf{a}' \; : \; \|\mathbf{c}_i - \mathbf{a}'\|^2 + \sum_{\mathbf{a} \in \pi_i} p(\mathbf{a}, \mathbf{a}') \leq \|\mathbf{c}_l - \mathbf{a}'\|^2 \right.$$

$$+ \sum_{a \in \pi_l} p(\mathbf{a}, \mathbf{a}') \text{ for each } l = 1, \ldots, k \Bigg\}. \tag{5.9}$$

We first show that the assignment step (i.e. Equation (5.9)) may lead to erroneous results.

Example 5.3.1 *Consider the one-dimensional dataset*

$$\mathcal{A} = \{\mathbf{a}_1, \mathbf{a}_2, \mathbf{a}_3, \mathbf{a}_4, \mathbf{a}_5\} = \{-2.9, \ -0.9, \ 0, \ 0.9, \ 2.9\}, \tag{5.10}$$

with $p(\mathbf{a}_i, \mathbf{a}_j) = p = 4$ when $i \neq j$. Consider the three-cluster partition

$$\Pi = \{\pi_1, \pi_2, \pi_3\}$$

with

$$\pi_1 = \{\mathbf{a}_1, \mathbf{a}_2\}, \ \pi_2 = \{\mathbf{a}_3\}, \ \pi_3 = \{\mathbf{a}_4, \mathbf{a}_5\}$$

(see Figure 5.1 *where the clusters are encircled). Note that*

$$Q(\Pi) = (2 + p) + 0 + (2 + p) = 4 + 2p = 12.$$

An application of the assignment step (5.9) leads to the three-cluster partition Π'

$$\pi_1' = \{\mathbf{a}_1\}, \ \pi_2' = \{\mathbf{a}_2, \mathbf{a}_3, \mathbf{a}_4\}, \ \pi_3' = \{\mathbf{a}_5\}$$

with

$$Q(\Pi') = 0 + (3p + 2(0.9)^2) + 0 = 1.62 + 3p = 13.62$$

(see Figure 5.2*).*

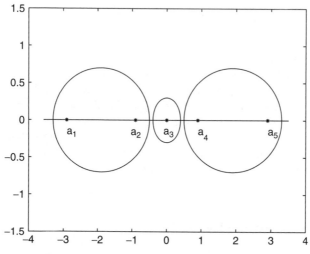

Figure 5.1   Initial three-cluster partition.

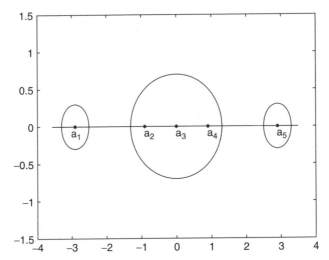

Figure 5.2   Three-cluster partition generated by batch k-means.

The assignment decision in Equation (5.9) ignores any anticipated change of centroid, and potential additional vector assignments coming from other clusters. As a result, the proposed batch iteration fails to improve the original partition (and, as the example shows, may lead to a partition of inferior quality).

Reassignment of a vector $\mathbf{a}$ from cluster π_i to cluster π_j changes the objective by

$$\Delta = \frac{|\pi_i|}{|\pi_i| - 1} \, ||\mathbf{c}(\pi_i) - \mathbf{a}||^2 - \frac{|\pi_j|}{|\pi_j| + 1} \, ||\mathbf{c}(\pi_j) - \mathbf{a}||^2$$
$$+ \sum_{\mathbf{a}' \in \pi_i} p(\mathbf{a}, \mathbf{a}') - \sum_{\mathbf{a}' \in \pi_j} p(\mathbf{a}, \mathbf{a}')$$

(see Equation (5.6)). We denote by $\Delta(\mathbf{a})$ the maximal value of the right hand side of Equation (5.11) over $j = 1, \ldots, m$. We note that removal of $\mathbf{a}$ from π_i and assigning it back to π_i is a reassignment with zero change of the objective. Hence $\Delta(\mathbf{a})$, the maximal value of the right hand side of Equation (5.11), is always nonnegative. To minimize the objective we shall select a vector $\mathbf{a}$ whose reassignment maximizes $\Delta(\mathbf{a})$. The incremental k-means algorithm we propose is given next. A single iteration of the algorithm applied to either one of the partitions Π or Π' of Example 5.3.1 generates a partition

$$\Pi'' = \{\{-2.9\}, \{-0.9, 0\}, \{0.9, 2.9\}\}$$

with $Q(\Pi'') = 10.405$ (see Figure 5.3).

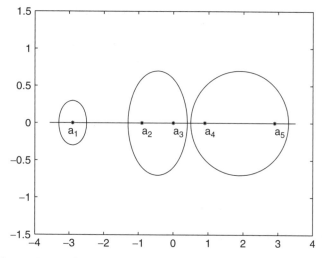

Figure 5.3   Optimal three-cluster partition generated by incremental k-means.

Algorithm 4 – Incremental k-means algorithms

1: For a user-supplied nonnegative tolerance `tol` ≥ 0 do the following:
2: Start with an initial partitioning

$$\Pi^{(0)} = \{\pi_1^{(0)}, \ldots, \pi_k^{(0)}\}.$$

3: Set the index of iteration $t = 0$.
4: Generate the partition `nextFV`$\left(\Pi^{(t)}\right)$.
5: **if** $\left[Q\left(\Pi^{(t)}\right) - Q\left(\text{nextFV}\left(\Pi^{(t)}\right)\right) > \text{tol}\right]$ **then**
6:      set $\Pi^{(t+1)} = \text{nextFV}\left(\Pi^{(t)}\right)$
7:      increment t by 1
8:      go to 5
9: **end if**
10: Stop.

5.3.2   Elimination of must-link constraints

We now reduce incremental k-means clustering of a vector set $\mathcal{A} = \{\mathbf{a}_1, \ldots, \mathbf{a}_m\}$ with both must-link and cannot-link constraints to that of (in general) a smaller vector set $\mathcal{B} = \{\mathbf{b}_1, \ldots, \mathbf{b}_M\}$ ($M \leq m$) with a different penalty function $P(\mathbf{b}, \mathbf{b}')$, and no must-link constraints. To simplify the presentation we assume throughout that $p(\mathbf{a}, \mathbf{a}') = 0$ for any pair of must-linked vectors $\mathbf{a}$ and $\mathbf{a}'$.

Consider the transitive closure of must-link constraints (see e.g. Basu et al. (2009)). For a vector $\mathbf{a} \in \mathcal{A}$ let $\pi(\mathbf{a})$ be a set of vectors $\mathbf{a}'$ in $\mathcal{A}$ so that there is a finite subset $\{\mathbf{a}_{i_1}, \ldots, \mathbf{a}_{i_p}\} \subseteq \mathcal{A}$ with $\mathbf{a} = \mathbf{a}_{i_1}$, $\mathbf{a}' = \mathbf{a}_{i_p}$, and $\mathbf{a}_{i_j}$ and $\mathbf{a}_{i_{j+1}}$ must-linked, $j = 1, \ldots, p - 1$. The sets $\pi(\mathbf{a})$ are equivalence classes, i.e.

1. for each $\mathbf{a}$, $\mathbf{a}' \in \mathcal{A}$ either $\pi(\mathbf{a}) = \pi(\mathbf{a}')$, or $\pi(\mathbf{a}) \cap \pi(\mathbf{a}') = \emptyset$;

2. $\mathcal{A} = \bigcup_{\mathbf{a} \in \mathcal{A}} \pi(\mathbf{a})$.

We denote the finite set collection $\{\pi(\mathbf{a})\}_{\mathbf{a} \in \mathcal{A}}$ by $\{\pi_1, \ldots, \pi_M\}$. For $i = 1, \ldots, M$ let

1. $\mathbf{b}_i = \mathbf{c}(\pi_i) = (1/|\pi_i|) \sum_{\mathbf{a} \in \pi_i} \mathbf{a}$, the centroid of π_i;

2. $q_i = Q(\pi_i)$, the quality of π_i;

3. $m(\mathbf{b}_i) = m_i = |\pi_i|$, the size of π_i.

The vector set $\mathcal{B} = \{\mathbf{b}_1, \ldots, \mathbf{b}_M\}$ is the new set to be clustered. For two vectors $\mathbf{b}_i$, $\mathbf{b}_j \in \mathcal{B}$ the penalty is defined by

$$P(\mathbf{b}_i, \mathbf{b}_j) = \sum_{\mathbf{a} \in \pi_i, \, \mathbf{a}' \in \pi_j} p(\mathbf{a}, \mathbf{a}'). \tag{5.11}$$

Each k-cluster partition $\Pi_\mathcal{B}$ of the set $\mathcal{B}$ induces a k-cluster partition $\Pi_\mathcal{A}$ of the set $\mathcal{A}$ with no must-link violations. We define the quality $Q_\mathcal{B}(\pi^\mathcal{B})$ of a subset $\pi^\mathcal{B} = \{\mathbf{b}_{i_1}, \ldots, \mathbf{b}_{i_p}\} \subseteq \mathcal{B}$ by

$$Q_\mathcal{B}(\pi^\mathcal{B}) = \sum_{j=1}^{p} m_{i_j} \|\mathbf{c} - \mathbf{b}_{i_j}\|^2 + \frac{1}{2} \sum_{l,j} P(\mathbf{b}_{i_l}, \mathbf{b}_{i_j}), \tag{5.12}$$

where

$$\mathbf{c} = \frac{m_{i_1} \mathbf{b}_{i_1} + \cdots + m_{i_p} \mathbf{b}_{i_p}}{m_{i_1} + \cdots + m_{i_p}}$$

is the (weighted) arithmetic mean of the set $\pi^\mathcal{B} = \{\mathbf{b}_{i_1}, \ldots, \mathbf{b}_{i_p}\}$ and the associated subset $\bigcup_{j=1}^{p} \pi_{i_j}$ of $\mathcal{A}$. The quality functions of the sets $\pi^\mathcal{B}$ and $\bigcup_{j=1}^{p} \pi_{i_j}$ are related as follows:

$$Q\left(\bigcup_{j=1}^{p} \pi_{i_j}\right) = \sum_{j=1}^{p} q_{i_j} + Q_\mathcal{B}(\pi^\mathcal{B}) \tag{5.13}$$

(for the unconstrained case see Kogan (2007a)). Hence, for each pair of associated partitions $\Pi_\mathcal{B}$ and $\Pi_\mathcal{A}$ the difference between $Q(\Pi_\mathcal{A})$ and $Q_\mathcal{B}(\Pi_\mathcal{B})$ is the same constant $\sum_{i=1}^{M} q_i$.

Incremental clustering of the set $\mathcal{B}$ is identical to Algorithm 4 with change Δ in the objective function caused by reassignment of a vector $\mathbf{b}$ from the cluster π_i^B to the cluster π_j^B given by

$$
\Delta = \frac{M_i \cdot m(\mathbf{b})}{M_i + m(\mathbf{b})} \left\| \mathbf{c}\left(\pi_i^B\right) - \mathbf{b} \right\|^2 - \frac{M_j \cdot m(\mathbf{b})}{M_j - m(\mathbf{b})} \left\| \mathbf{c}\left(\pi_j^B\right) - \mathbf{b} \right\|^2
$$
$$
+ \sum_{\mathbf{b}' \in \pi_i^B} P(\mathbf{b}, \mathbf{b}') - \sum_{\mathbf{b}' \in \pi_j^B} P(\mathbf{b}, \mathbf{b}'), \tag{5.14}
$$

where $M_l = \sum_{\mathbf{b} \in \pi_l^B} m(\mathbf{b})$. In what follows, we extend these results to Bregman distances.

5.3.3 Clustering with Bregman divergences

Let $\psi : \mathbf{R}^n \to (-\infty, +\infty]$ be a closed proper convex function (Rockafellar 1970). Suppose that ψ is continuously differentiable on $\mathrm{int}(\mathrm{dom}\ \psi) \neq \emptyset$. The Bregman distance (also called 'Bregman divergence') $D_\psi : \mathrm{dom}\ \psi \times \mathrm{int}(\mathrm{dom}\ \psi) \to \mathbf{R}_+$ is defined by

$$
D_\psi(\mathbf{x}, \mathbf{y}) = \psi(\mathbf{x}) - \psi(\mathbf{y}) - \nabla\psi(\mathbf{y})(\mathbf{x} - \mathbf{y}), \tag{5.15}
$$

where $\nabla\psi$ is the gradient of ψ.

This function measures the convexity of ψ, i.e. $D_\psi(\mathbf{x}, \mathbf{y}) \geq 0$, if and only if the gradient inequality for ψ holds, i.e. if and only if ψ is convex. With ψ strictly convex one has $D_\psi(\mathbf{x}, \mathbf{y}) \geq 0$ and $D_\psi(\mathbf{x}, \mathbf{y}) = 0$ iff $\mathbf{x} = \mathbf{y}$.

Note that $D_\psi(\mathbf{x}, \mathbf{y})$ is not a distance (it is, in general, not symmetric and does not satisfy the triangle inequality). With $\psi(\mathbf{x}) = ||\mathbf{x}||^2$ ($\mathrm{dom}\ \psi = \mathbf{R}^n$) one has $D_\psi(\mathbf{x}, \mathbf{y}) = ||\mathbf{x} - \mathbf{y}||^2$. With $\psi(\mathbf{x}) = \sum_{j=1}^n \mathbf{x}[j] \log \mathbf{x}[j] - \mathbf{x}[j]$ ($\mathrm{dom}\ \psi = \mathbf{R}_+^n$ with the convention $0 \log 0 = 0$), we obtain the Kullback–Leibler relative entropy distance

$$
D_\psi(\mathbf{x}, \mathbf{y}) = \sum_{j=1}^n \mathbf{x}[j] \log \frac{\mathbf{x}[j]}{\mathbf{y}[j]} + \mathbf{y}[j] - \mathbf{x}[j] \ \forall\ (\mathbf{x}, \mathbf{y}) \in \mathbf{R}_+^n \times \mathbf{R}_{++}^n. \tag{5.16}
$$

Note that under the additional assumption $\sum_{j=1}^n \mathbf{x}[j] = \sum_{j=1}^n \mathbf{y}[j] = 1$, the Bregman divergence $D_\psi(\mathbf{x}, \mathbf{y})$ reduces to $\sum_{j=1}^n \mathbf{x}[j] \log(\mathbf{x}[j]/\mathbf{y}[j])$ (for additional examples of Bregman distances see e.g. Banerjee et al. (2005) and Teboulle et al. (2006)). Note that Bregman distance $D_\psi(\mathbf{x}, \mathbf{y})$ is convex with respect to the $\mathbf{x}$ variable. Hence, centroid computation in Equation (5.1) is an 'easy' optimization problem.

By reversing the order of variables in D_ψ, i.e.

$$
\overleftarrow{D}_\psi(\mathbf{x}, \mathbf{y}) = D_\psi(\mathbf{y}, \mathbf{x}) = \psi(\mathbf{y}) - \psi(\mathbf{x}) - \nabla\psi(\mathbf{x})(\mathbf{y} - \mathbf{x}) \tag{5.17}
$$

(compare with Equation (5.15)) and using the kernel

$$\psi(\mathbf{x}) = \frac{\nu}{2}\|\mathbf{x}\|^2 + \mu \left[\sum_{j=1}^{n} \mathbf{x}[j] \log \mathbf{x}[j] - \mathbf{x}[j] \right], \qquad (5.18)$$

we obtain

$$\overleftarrow{D_\psi}(\mathbf{x}, \mathbf{y}) = D_\psi(\mathbf{y}, \mathbf{x}) = \frac{\nu}{2}\|\mathbf{y} - \mathbf{x}\|^2 + \mu \sum_{j=1}^{n} \left[\mathbf{y}[j] \log \frac{\mathbf{y}[j]}{\mathbf{x}[j]} + \mathbf{x}[j] - \mathbf{y}[j] \right].$$

$$(5.19)$$

While in general $\overleftarrow{D_\psi}(\mathbf{x}, \mathbf{y})$ given by Equation (5.16) is not necessarily convex in $\mathbf{x}$, when $\psi(\mathbf{x})$ is given either by $\|\mathbf{x}\|^2$ or by $\sum_{j=1}^{n} \mathbf{x}[j] \log \mathbf{x}[j] - \mathbf{x}[j]$ the resulting functions $\overleftarrow{D_\psi}(\mathbf{x}, \mathbf{y})$ are strictly convex with respect to the first variable.

Extension of Algorithm 4 to 'reversed' Bregman distances requires the following:

1. The ability to compute $\mathbf{c}(\pi)$ for a finite set π (see Equation (5.1)).

2. A convenient expression for $Q_B(\pi^B)$ of a subset $\pi^B = \{\mathbf{b}_{i_1}, \ldots, \mathbf{b}_{i_p}\} \subseteq B$ (see (5.12)).

3. A convenient formula for the change Δ in the objective function caused by reassignment of a vector $\mathbf{b}$ from the cluster π_i^B to the cluster π_j^B (see (5.14)).

We next list results already available in the literature and relevant to the above three points. The first result[1] holds for all Bregman divergences with reversed order of variables $\overleftarrow{D_\psi}(\mathbf{x}, \mathbf{y}) = D_\psi(\mathbf{y}, \mathbf{x})$ (see Banerjee et al. (2005)):

Theorem 5.3.2 *If* $\mathbf{z} = (\mathbf{a}_1 + \cdots + \mathbf{a}_m)/m$, *then* $\sum_{i=1}^{m} D_\psi(\mathbf{a}_i, \mathbf{z}) \leq \sum_{i=1}^{m} D_\psi (\mathbf{a}_i, \mathbf{x})$.

The result shows that the centroid of any set equipped with reversed Bregman distance is given by the arithmetic mean.

The change Δ in the objective Q caused by moving a vector $\mathbf{a}$ from cluster π_i to cluster π_j is given by

$$\Delta = (m_i - 1)[\psi(\mathbf{c}_i^-) - \psi(\mathbf{c}_i)] - \psi(\mathbf{c}_i) + (m_j + 1)[\psi(\mathbf{c}_j^+) - \psi(\mathbf{c}_j)] + \psi(\mathbf{c}_j),$$

$$(5.20)$$

where m_i and m_j denote the size of the clusters π_i and π_j, $\mathbf{c}_i^-$ is the centroid of π_i with $\mathbf{a}$ being removed, and $\mathbf{c}_j^+$ is the centroid of π_j with $\mathbf{a}$ being added (see Kogan (2007a)).

---

[1] Note that this distance-like function is not necessarily convex with respect to $\mathbf{x}$.

In text mining applications, due to sparsity of the data vector $\mathbf{a}$, most coordinates of centroids $\mathbf{c}^-$, $\mathbf{c}^+$, and $\mathbf{c}$ coincide. Hence, when the function ψ is separable, computation of $\psi(\mathbf{c}_i^-)$ and $\psi(\mathbf{c}_j^+)$ is relatively cheap.

Elimination of must-links requires an analogue of Equations (5.12) and (5.14). The following two statements are provided by Kogan (2007a):

Theorem 5.3.3 *If* $\mathcal{A} = \pi_1 \cup \pi_2 \cup \cdots \cup \pi_k$ *with* $m_i = |\pi_i|$, $\mathbf{c}_i = \mathbf{c}(\pi_i)$, $i = 1, \ldots, k$,

$$\mathbf{c} = \mathbf{c}(\mathcal{A}) = \frac{m_1}{m}\mathbf{c}_1 + \cdots + \frac{m_k}{m}\mathbf{c}_k, \text{ where } m = m_1 + \cdots + m_k,$$

and $\Pi = \{\pi_1\,\pi_2, \ldots, \pi_k\}$, *then*

$$Q(\Pi) = \sum_{i=1}^{k} Q(\pi_i) + \sum_{i=1}^{k} m_i d(\mathbf{c}, \mathbf{c}_i) = \sum_{i=1}^{k} Q(\pi_i) + \sum_{i=1}^{k} m_i [\psi(\mathbf{c}_i) - \psi(\mathbf{c})].$$

$$(5.21)$$

Theorem 5.3.4 *Let* $\Pi_{\mathcal{B}} = \{\pi_1^{\mathcal{B}}, \ldots, \pi_k^{\mathcal{B}}\}$ *be a* k-*cluster partition of the set* $\mathcal{B} = \{\mathbf{b}_1, \ldots, \mathbf{b}_M\}$. *If* $\Pi_{\mathcal{B}}'$ *is a partition obtained from* $\mathcal{B}$ *by removal of a single vector* $\mathbf{b}$ *from cluster* $\pi_i^{\mathcal{B}}$ *with centroid* $\mathbf{c}_i = \mathbf{c}\left(\pi_i^{\mathcal{B}}\right)$ *and assignment of* $\mathbf{b}$ *to* $\pi_j^{\mathcal{B}}$ *with centroid* $\mathbf{c}_j = \mathbf{c}\left(\pi_j^{\mathcal{B}}\right)$, *then the change of quality* $\Delta = Q_{\mathcal{B}}(\Pi_{\mathcal{B}}) - Q_{\mathcal{B}}(\Pi_{\mathcal{B}}')$ *is given by*

$$\Delta = [M_i - m(\mathbf{b})]\left[\psi(\mathbf{c}_i^-) - \psi(\mathbf{c}_i)\right] - m(\mathbf{b})\psi(\mathbf{c}_i)$$

$$+ \left[M_j + m(\mathbf{b})\right]\left[\psi(\mathbf{c}_j^+) - \psi(\mathbf{c}_j)\right] + m(\mathbf{b})\psi(\mathbf{c}_j). \quad (5.22)$$

We are now in a position to present the constrained clustering algorithm for a dataset with 'reversed' Bregman distance (see Algorithm 5). The next section describes a constrained clustering algorithm based on a nonlinear optimization approach.

Algorithm 5 – Constrained k-means with Bregman distance

1: For a dataset $\mathcal{A}$, a set of *must-link* and *cannot-link* constraints, and a user-supplied nonnegative tolerance `tol` ≥ 0 do the following:

2: Substitute cannot-link constraints by a penalty function p.

3: Build a transitive closure $\mathcal{B} = \{\mathbf{b}_1, \ldots, \mathbf{b}_M\}$ of must-link constraints.

4: Use Equation (5.11) to define the penalty $P(\mathbf{b}_i, \mathbf{b}_j)$ for each pair $\mathbf{b}_i, \mathbf{b}_j \in \mathcal{B}$.

5: Start with an initial k-cluster partitioning $\Pi_{\mathcal{B}}^{(0)} = \{\pi_1^{\mathcal{B}}, \ldots, \pi_k^{\mathcal{B}}\}$.

6: Set the index of iteration $t = 0$.

7: Use the change of quality

$$\Delta = [M_i - m(\mathbf{b})]\left[\psi(\mathbf{c}_i^-) - \psi(\mathbf{c}_i)\right] - m(\mathbf{b})\psi(\mathbf{c}_i)$$

$$+ \left[M_j + m(\mathbf{b}) \right] \left[\psi(\mathbf{c}_j^+) - \psi(\mathbf{c}_j) \right] + m(\mathbf{b})\psi(\mathbf{c}_j)$$
$$+ \sum_{\mathbf{b}' \in \pi_i^B} P(\mathbf{b}, \mathbf{b}') - \sum_{\mathbf{b}' \in \pi_j^B} P(\mathbf{b}, \mathbf{b}')$$

generated by removal of a single vector $\mathbf{b}$ from cluster π_i^B with centroid $\mathbf{c}_i = \mathbf{c}\left(\pi_i^B\right)$ and assignment of $\mathbf{b}$ to π_j^B with centroid $\mathbf{c}_j = \mathbf{c}\left(\pi_j^B\right)$ to identify the partition $\texttt{nextFV}\left(\Pi^{(t)}\right)$.

8: **if** $\left[Q\left(\Pi^{(t)}\right) - Q\left(\texttt{nextFV}\left(\Pi^{(t)}\right)\right) > \texttt{tol}\right]$ **then**
9:      set $\Pi^{(t+1)} = \texttt{nextFV}\left(\Pi^{(t)}\right)$
10:     increment t by 1
11:     go to 8
12: **end if**
13: Stop.

5.4   Constrained smoka type clustering

First we briefly recall smoka type clustering (Teboulle and Kogan 2005). Note that for a vector $\mathbf{a}$ and k vectors $\mathbf{x}_1, \ldots, \mathbf{x}_k$ one has

$$\lim_{s \to 0} -s \log \left(\sum_{l=1}^{k} e^{-\frac{\|\mathbf{x}_l - \mathbf{a}\|^2}{s}} \right) = \min \left\{ \|\mathbf{x}_1 - \mathbf{a}\|^2, \ldots, \|\mathbf{x}_k - \mathbf{a}\|^2 \right\}. \qquad (5.23)$$

When $\mathbf{x}_1, \ldots, \mathbf{x}_k$ are centroids of a k-cluster partition $\Pi = \{\pi_1, \ldots, \pi_k\}$ one has

$$Q(\Pi) = \sum_{i=1}^{k} \sum_{\mathbf{a} \in \pi_i} \|\mathbf{x}_i - \mathbf{a}\| = \sum_{\mathbf{a} \in \mathcal{A}} \min \left\{ \|\mathbf{x}_1 - \mathbf{a}\|^2, \ldots, \|\mathbf{x}_k - \mathbf{a}\|^2 \right\}$$

$$= \lim_{s \to 0} \sum_{\mathbf{a} \in \mathcal{A}} \left[-s \log \left(\sum_{l=1}^{k} e^{-\frac{\|\mathbf{x}_l - \mathbf{a}\|^2}{s}} \right) \right]. \qquad (5.24)$$

The right hand side of Equation (5.24) shows that the problem of finding the best k-cluster partition with no constraints can be restated as the problem of identifying the k best centroids $\mathbf{x}_1, \ldots, \mathbf{x}_k$. While both expressions

$$\sum_{\mathbf{a} \in \mathcal{A}} \left[-s \log \left(\sum_{l=1}^{k} e^{-\frac{\|\mathbf{x}_l - \mathbf{a}\|^2}{s}} \right) \right] \quad \text{and} \quad \sum_{\mathbf{a} \in \mathcal{A}} \min \left\{ \|\mathbf{x}_1 - \mathbf{a}\|^2, \ldots, \|\mathbf{x}_k - \mathbf{a}\|^2 \right\}$$

are functions of $\mathbf{x}_1, \ldots, \mathbf{x}_k$, the one on the left is differentiable, while the one on the right is not. This observation suggests use of the smooth approximation

$$\sum_{\mathbf{a} \in \mathcal{A}} \left[-s \log \left(\sum_{l=1}^{k} e^{-\frac{\|\mathbf{x}_l - \mathbf{a}\|^2}{s}} \right) \right]$$

in order to approximate optimal centroids. Application of smooth approximations to k-means clustering appears, for example, in Rose et al. (1990), Marroquin and Girosi (1993), Nasraoui and Krishnapuram (1995), Teboulle and Kogan (2005), and Teboulle (2007).

Next we briefly describe smoka clustering with cannot-link constraints only. For two vectors $\mathbf{a}$, $\mathbf{a}'$, and a set of k vectors $\mathbf{x}_1, \ldots, \mathbf{x}_k$, one has

$$\lim_{s \to 0} -s \log \left(\sum_{i,j=1}^{k} e^{-\frac{\|\mathbf{x}_i - \mathbf{a}\|^2 + \|\mathbf{x}_j - \mathbf{a}'\|^2}{s}} \right) = \min_{i,j} \left\{ \|\mathbf{x}_i - \mathbf{a}\|^2 + \|\mathbf{x}_j - \mathbf{a}'\|^2 \right\}. \quad (5.25)$$

We denote the left hand side of (5.25) by $\psi(\mathbf{a}, \mathbf{a}')$, and define $\phi(\mathbf{a}, \mathbf{a}')$ as

$$\lim_{s \to 0} -s \log \left(\sum_{i=1}^{k} e^{-\frac{\|\mathbf{x}_i - \mathbf{a}\|^2 + \|\mathbf{x}_i - \mathbf{a}'\|^2}{s}} \right) = \min_{i} \left\{ \|\mathbf{x}_i - \mathbf{a}\|^2 + \|\mathbf{x}_i - \mathbf{a}'\|^2 \right\}. \quad (5.26)$$

Clearly $\psi(\mathbf{a}, \mathbf{a}') \leq \phi(\mathbf{a}, \mathbf{a}')$, and the equality holds only when $\mathbf{a}$ and $\mathbf{a}'$ belong to the same cluster. This observation motivates the introduction of a penalty function for cannot-linked vectors $\mathbf{a}, \mathbf{a}'$ as $p(\mathbf{a}, \mathbf{a}') = \rho \left(\phi(\mathbf{a}, \mathbf{a}') - \psi(\mathbf{a}, \mathbf{a}') \right)$ where $\rho : \mathbf{R}_+ \to \mathbf{R}_+$ is a monotonically increasing function with $\rho(0) = 0$ so that $p(\mathbf{a}, \mathbf{a}') = 0$ when $\mathbf{a}$ and $\mathbf{a}'$ belong to the same cluster (the simplest but, perhaps, not the best choice for the function ρ is $\rho(t) = t$).

Since we intend to approximate the right hand side of Equations (5.25) and (5.26) by the corresponding expressions on the left hand side with 'small' values of s, we shall consider penalty function $p_s(\mathbf{a}, \mathbf{a}') = \rho \left(\phi_s(\mathbf{a}, \mathbf{a}') - \psi_s(\mathbf{a}, \mathbf{a}') \right)$ where

$$\psi_s(\mathbf{a}, \mathbf{a}') = -s \log \left(\sum_{i,j=1}^{k} e^{-\frac{\|\mathbf{x}_i - \mathbf{a}\|^2 + \|\mathbf{x}_j - \mathbf{a}'\|^2}{s}} \right) \quad (5.27)$$

and

$$\phi_s(\mathbf{a}, \mathbf{a}') = -s \log \left(\sum_{i=1}^{k} e^{-\frac{\|\mathbf{x}_i - \mathbf{a}\|^2 + \|\mathbf{x}_i - \mathbf{a}'\|^2}{s}} \right). \quad (5.28)$$

For fixed vectors $\mathbf{a}$, $\mathbf{a}'$ the expressions ψ_s and ϕ_s are functions of $\mathbf{x} = (\mathbf{x}_1^T, \ldots, \mathbf{x}_k^T)^T \in \mathbf{R}^{kn}$, and we shall abuse notation and denote the penalty by $p_s(\mathbf{x}; \mathbf{a}, \mathbf{a}')$.

Our goal is to minimize

$$F_s(\mathbf{x}) = \sum_{i=1}^{m} -s \log \left(\sum_{l=1}^{k} e^{-\frac{\|\mathbf{x}_l - \mathbf{a}_i\|^2}{s}} \right) + \frac{1}{2} \sum_{\mathbf{a}, \mathbf{a}' \in \mathcal{A}} p_s(\mathbf{x}; \mathbf{a}, \mathbf{a}') \quad (5.29)$$

with respect to $\mathbf{x} \in \mathbf{R}^{kn}$.

We now turn to must-link constraints. Elimination of must-link constraints is again based on 'collapsing' a set of vectors that should be placed together in the same cluster into the set's centroid $\mathbf{b}$ and clustering the transitive closure of must-link constraints $\mathcal{B} = \{\mathbf{b}_1, \ldots, \mathbf{b}_M\}$. This approach with no cannot-link constraints was introduced in Kogan (2007b). The objective function to be minimized is

$$-s \sum_{i=1}^{M} m_i \log \left(\sum_{l=1}^{k} e^{-\frac{\|\mathbf{x}_l - \mathbf{b}_i\|^2}{s}} \right),$$ (5.30)

where $m_i = m(b_i)$. To incorporate cannot-link constraints we again introduce a penalty function. The penalty $P_s(\mathbf{x}; \mathbf{b}, \mathbf{b}')$ should reflect the cluster size $m(\mathbf{b})$ and is defined as follows:

$$P_s(\mathbf{x}; \mathbf{b}, \mathbf{b}') = \left[m(\mathbf{b}) + m(\mathbf{b}') \right]$$

$$\times \rho \left\{ -s \left[\log \left(\sum_{i=1}^{k} e^{-\frac{\|\mathbf{x}_i - \mathbf{b}\|^2 + \|\mathbf{x}_i - \mathbf{b}'\|^2}{s}} \right) - \log \left(\sum_{i,j=1}^{k} e^{-\frac{\|\mathbf{x}_i - \mathbf{b}\|^2 + \|\mathbf{x}_j - \mathbf{b}'\|^2}{s}} \right) \right] \right\}.$$ (5.31)

We shall abuse notation and denote the objective to be minimized by $F_s(\mathbf{x})$:

$$F_s(\mathbf{x}) = -s \sum_{i=1}^{M} m_i \log \left(\sum_{l=1}^{k} e^{-\frac{\|\mathbf{x}_l - \mathbf{b}_i\|^2}{s}} \right) + \frac{1}{2} \sum_{\mathbf{b}, \mathbf{b}' \in \mathcal{B}} P_s(\mathbf{x}; \mathbf{b}, \mathbf{b}'),$$ (5.32)

where $\mathbf{x} = \left(\mathbf{x}_1^T, \ldots, \mathbf{x}_k^T \right)^T$. The clustering algorithm is presented next (see Algorithm 6). The following section describes the constrained clustering algorithm designed to handle unit length vectors.

Algorithm 6 – Constrained smoka clustering

1: For a dataset $\mathcal{A}$, a set of *must-link* and *cannot-link* constraints, positive parameters s and ϵ, and a user-supplied nonnegative tolerance tol ≥ 0 do the following:
2: Build a transitive closure $\mathcal{B} = \{\mathbf{b}_1, \ldots, \mathbf{b}_M\}$ of must-link constraints.
3: Select initial cluster set $\mathbf{x}^0 \in \mathbf{R}^{kn}$ and set the index of iterations $t = 0$.
4: Use gradient descent to generate $\mathbf{y}$ from $\mathbf{x}^{(t)}$.
5: **if** $F_s\left(\mathbf{x}^{(t)}\right) - F_s(\mathbf{y}) >$ tol **then**
6:     increment t by 1
7:     set $\mathbf{x}^{(t)} = \mathbf{y}$
8:     go to 5
9: **end if**
10: Stop.

5.5   Constrained spherical k-means

This section describes a clustering algorithm designed to handle l_2 unit norm vectors. The unconstrained version of the algorithm introduced in Dhillon and Modha (1999) was motivated by information retrieval (IR) applications and designed to handle vectors with nonnegative entries. In Dhillon et al. (2003) the algorithm was extended to vector datasets with arbitrary entries (see also Kogan (2007a) for detailed treatment of general n-dimensional datasets).

The algorithm is reminiscent of the quadratic k-means algorithm, but the 'distance' between two unit vectors $\mathbf{x}$ and $\mathbf{y}$ is measured by $d(\mathbf{x}, \mathbf{y}) = \mathbf{x}^T \mathbf{y}$ (so that the two unit vectors $\mathbf{x}$ and $\mathbf{y}$ are equal if and only if $d(\mathbf{x}, \mathbf{y}) = 1$). We define the set $\mathcal{C}$ housing centroids as the union of the unit $(n-1)$-dimensional l_2 sphere

$$\mathcal{S}_2^{n-1} = \{\mathbf{x} \ : \ \mathbf{x} \in \mathbf{R}^n, \ \mathbf{x}^T \mathbf{x} = 1\}$$

centered at the origin (when it does not lead to ambiguity we shall denote the sphere just by $\mathcal{S}$).

For a set of vectors $\mathcal{A} = \{\mathbf{a}_1, \ldots, \mathbf{a}_m\} \subset \mathbf{R}^n$, and the 'distance-like' function $d(\mathbf{x}, \mathbf{a}) = \mathbf{a}^T \mathbf{x}$, we define centroid $\mathbf{c} = \mathbf{c}(\mathcal{A})$ of the set $\mathcal{A}$ as a solution of the *maximization* problem

$$\mathbf{c} = \begin{cases} \arg\max \left\{ \sum_{\mathbf{a} \in \mathcal{A}} \mathbf{x}^T \mathbf{a}, \ \mathbf{x} \in \mathcal{S} \right\} & \text{if} & \mathbf{a}_1 + \cdots + \mathbf{a}_m \neq 0, \\ 0 & \text{otherwise.} \end{cases} \tag{5.33}$$

Equation (5.33) immediately yields

$$\mathbf{c}(\mathcal{A}) = \begin{cases} \dfrac{\mathbf{a}_1 + \cdots + \mathbf{a}_m}{\|\mathbf{a}_1 + \cdots + \mathbf{a}_m\|} & \text{if} & \mathbf{a}_1 + \cdots + \mathbf{a}_m \neq 0, \\ 0 & \text{otherwise.} \end{cases} \tag{5.34}$$

Note that:

1. For $\mathcal{A} \subset \mathbf{R}_+^n$ (which is typical for many IR applications) the sum of the vectors in $\mathcal{A}$ is never zero, and $\mathbf{c}(\mathcal{A})$ is a unit length vector.

2. The quality of the set $\mathcal{A}$ is just $Q(\mathcal{A}) = \sum_{\mathbf{a} \in \mathcal{A}} \mathbf{a}^T \mathbf{c}(\mathcal{A}) = \|\mathbf{a}_1 + \cdots + \mathbf{a}_m\|$.

3. While the motivation for spherical k-means is provided by IR applications dealing with vectors with nonnegative coordinates residing on the unit sphere, Equation (5.34) provides solutions to the maximization problem in Equation (5.33) for *any* set $\mathcal{A} \subset \mathbf{R}^n$.

Spherical batch k-means is a procedure similar to the batch k-means algorithm with the obvious substitution of min by max in Equation (5.4).

5.5.1   Spherical k-means with cannot-link constraints only

In the presence of cannot-link constraints we introduce a *nonpositive* symmetric penalty function $p(\mathbf{a}, \mathbf{a}') \leq 0$. For a cluster π we define

$$Q(\pi) = \sum_{\mathbf{a} \in \pi} \mathbf{a}^T \mathbf{c}(\pi) + \frac{1}{2} \sum_{\mathbf{a}, \mathbf{a}' \in \pi} p(\mathbf{a}, \mathbf{a}'), \qquad (5.35)$$

with $c(\pi)$ given by Equation (5.34). The quality of partition $\Pi = \{\pi_1, \cdots, \pi_k\}$ is defined by

$$Q(\Pi) = \sum_{i=1}^{k} Q(\pi_i). \qquad (5.36)$$

We first show that a straightforward adaptation of spherical batch k-means to datasets equipped with a penalty may lead to erroneous results.

Example 5.5.1 *Let* $\mathcal{A} = \{\mathbf{a}_1, \mathbf{a}_2, \mathbf{a}_3, \mathbf{a}_4, \mathbf{a}_5\} \subset \mathbf{R}^2$ *with*

$$\mathbf{a}_1 = \begin{bmatrix} 1 \\ 0 \end{bmatrix}, \ \mathbf{a}_2 = \begin{bmatrix} \cos 31° \\ \sin 31° \end{bmatrix}, \ \mathbf{a}_3 = \begin{bmatrix} \cos 45° \\ \sin 45° \end{bmatrix},$$

$$\mathbf{a}_4 = \begin{bmatrix} \cos 59° \\ \sin 59° \end{bmatrix}, \ \mathbf{a}_5 = \begin{bmatrix} 0 \\ 1 \end{bmatrix},$$

and $p(\mathbf{a}_i, \mathbf{a}_j) = -1$, $i \neq j$. *Consider an initial three-cluster partition (see Figure 5.4)*

$$\Pi = \{\pi_1, \pi_2, \pi_3\}, \ \text{with} \ \pi_1 = \{\mathbf{a}_1, \mathbf{a}_2\}, \pi_2 = \{\mathbf{a}_3\}, \pi_3 = \{\mathbf{a}_4, \mathbf{a}_5\},$$

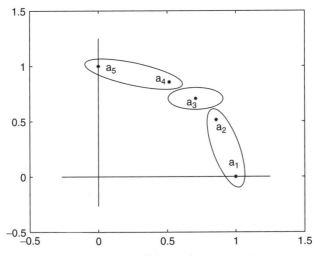

Figure 5.4   Initial three-cluster partition.

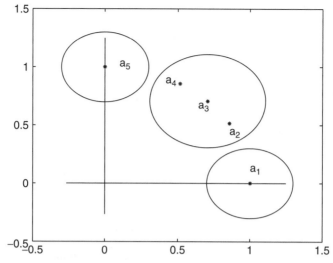

Figure 5.5   Three-cluster partition generated by spherical batch k-means.

with $Q(\Pi) = 4.8546 + 2p$. *An application of one batch iteration generates partition* Π' *with* $Q(\Pi') = 4.9406 + 3p$ *(see Figure 5.5). For each penalty* $p < -0.086$ *one has* $Q(\Pi') < Q(\Pi)$, *i.e. an application of one iteration of the algorithm leads to an inferior partition.*

The incremental version of spherical batch *k*-means is analogous to that of *k*-means with the obvious reverse of the inequality in Equation (5.7). An application of a single iteration of incremental algorithm to partition Π (see Example 5.5.1) generates partition $\Pi'' = \{\{\mathbf{a}_1, \mathbf{a}_2\}, \{\mathbf{a}_3, \mathbf{a}_4\}, \{\mathbf{a}_5\}\}$ with $Q(\Pi'') = 4.9124 + 2p < 4.8546 + 2p = Q(\Pi)$ (see Figure 5.6).

Algorithm 7 – Incremental spherical *k*-means algorithm

1: Given user-supplied tolerance $\mathtt{tol_I} \geq 0$, do the following:
2: Start with a partitioning $\Pi^{(0)}$.
3: Set the index of iteration $t = 0$.
4: Generate $\mathtt{nextFV}\left(\Pi^{(t)}\right)$.
5: **if** $\left[Q\left(\mathtt{nextFV}\left(\Pi^{(t)}\right)\right) - Q\left(\Pi^{(t)}\right) > \mathtt{tol_I}\right]$ **then**
6:    set $\Pi^{(t+1)} = \mathtt{nextFV}\left(\Pi^{(t)}\right)$
7:    increment t by 1
8:    go to 5
9: **end if**
10: Stop.

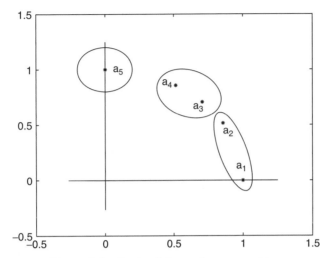

Figure 5.6   Optimal three-cluster partition.

5.5.2   Spherical k-means with cannot-link and must-link constraints

We start with an elementary observation. If $\pi = \{\mathbf{a}_1, \cdots, \mathbf{a}_m\}$ and $\pi' = \{\mathbf{a}'_1, \ldots, \mathbf{a}'_{m'}\}$ are two clusters, then

$$Q(\pi \cup \pi') = \left| \sum_{\mathbf{a} \in \pi} \mathbf{a} + \sum_{\mathbf{a}' \in \pi'} \mathbf{a}' \right| + \sum_{\mathbf{a} \in \pi, \, \mathbf{a}' \in \pi'} p(\mathbf{a}, \mathbf{a}'). \tag{5.37}$$

By setting $\mathbf{b} = \sum_{\mathbf{a} \in \pi} \mathbf{a}$, $\mathbf{b}' = \sum_{\mathbf{a}' \in \pi'} \mathbf{a}'$, and $P(\mathbf{b}, \mathbf{b}') = \sum_{\mathbf{a} \in \pi, \, \mathbf{a}' \in \pi'} p(\mathbf{a}, \mathbf{a}')$ one gets

$$Q(\pi \cup \pi') = |\mathbf{b} + \mathbf{b}'| + P(\mathbf{b}, \mathbf{b}').$$

This observation makes repetition of the construction presented in Section 5.3.2 possible. Consider the transitive closure $\{\pi_1, \ldots, \pi_M\}$ of must-link constraints. For $i = 1, \ldots, M$ let $\mathbf{b}_i = \sum_{\mathbf{a} \in \pi_i} \mathbf{a}$, and for each pair of indices $1 \le i, j \le M$ denote $\sum_{\mathbf{a} \in \pi_i, \, \mathbf{a}' \in \pi_j} p(\mathbf{a}, \mathbf{a}')$ by $P(\mathbf{b}_i, \mathbf{b}_j)$.

Our goal now is to cluster the set $\mathcal{B} = \{\mathbf{b}_1, \ldots, \mathbf{b}_M\}$. For a subset $\pi^{\mathcal{B}} = \{\mathbf{b}_{i_1}, \ldots, \mathbf{b}_{i_p}\} \subseteq \mathcal{B}$ the quality of the set is denoted by $Q_{\mathcal{B}}(\pi^{\mathcal{B}})$ and is defined by

$$Q_{\mathcal{B}}(\pi^{\mathcal{B}}) = \left| \sum_{\mathbf{b} \in \pi^{\mathcal{B}}} \mathbf{b} \right| + \frac{1}{2} \sum_{\mathbf{b}, \, \mathbf{b}' \in \pi^{\mathcal{B}}} P(\mathbf{b}, \mathbf{b}'). \tag{5.38}$$

The quality of the set $\pi^{\mathcal{B}}$ is equal to the quality of the associated subset $\pi^{\mathcal{A}} = \bigcup_{j=1}^{p} \pi_{i_j}$ of $\mathcal{A}$, i.e. $Q_{\mathcal{B}}\left(\pi^{\mathcal{B}}\right) = Q\left(\pi^{\mathcal{A}}\right)$.

The incremental spherical k-means algorithm for the dataset $\mathcal{A}$ with penalty function p and must-link constraints is identical to Algorithm 7 applied to the dataset $\mathcal{B}$ equipped with penalty function P and with no must-link constraints.

5.6 Numerical experiments

We now demonstrate that useful cannot-link constraints may lead to superior clustering results. We apply Algorithm 4 to a small three collection dataset classic3:[2]

- DC0 (Medlars Collection 1033 medical abstracts)

- DC1 (CISI Collection 1460 information science abstracts)

- DC2 (Cranfield Collection 1398 aerodynamics abstracts).

We denote the overall collection of 3891 documents by DC. Many clustering algorithms are capable of partitioning DC into three clusters with small (but not zero) 'misclassification' (see e.g. Dhillon et al. (2003); Dhillon and Modha (2001)).

We preprocess all the text datasets following the methodology of Dhillon et al. (2003), so that the clustering algorithm deals with 3891 vectors of dimension 600. An application of PDDP (Principal Direction Divisive Partitioning; see Boley (1998)) generates the initial three-cluster partition for DC. The confusion matrix for the partition is given in Table 5.1. This partition is used later as an input for both Algorithm 4 and Algorithm 7. Both algorithms are applied to the dataset with no must-link constraints. The penalty function $p(\mathbf{a}, \mathbf{a}')$ is defined as follows. For collection DC0 we sort all the document vectors $\mathbf{a}_{00}, \mathbf{a}_{01}, \mathbf{a}_{02} \ldots$ with respect to the distance to the collection average ($\mathbf{a}_{00}$ is the nearest). We select first r_0 vectors $\mathbf{a}_{00}, \mathbf{a}_{01}, \ldots \mathbf{a}_{0r_0-1}$ and for each $\mathbf{a}$ not in D0 define $p(\mathbf{a}_{0i}, \mathbf{a}) = p > 0$, $i = 1, \ldots, r_0 - 1$. For the other two document collections DC1 and DC2 the penalty function is defined analogously.

Table 5.1   PDDP generated 'confusion' matrix with **250** 'misclassified' documents.

Cluster/DocCol	DC0	DC1	DC2
Cluster 0	1362	13	6
Cluster 1	7	1372	120
Cluster 2	91	13	907

---

[2] Available from http://www.cs.utk.edu/~lsi.

5.6.1   Quadratic k-means

An application of Algorithm 4 with zero penalty and `tol` $= 0.001$ (i.e. just incremental k-means) to the PDDP generated partition improves the confusion matrix (see Table 5.2). Algorithm 4 with $p = 0.01$ generates the final partition with the confusion matrix given in Table 5.3. The penalty increase to 0.09 leads to the perfect diagonal confusion matrix given in Table 5.4. The values of penalty versus 'misclassification' of final partitions generated by Algorithm 4 with `tol` $= 0.001$ are given in Table 5.5. In these experiments $r_0 = r_1 = r_2 = 1$. Selection of $r_0 = r_1 = r_2 = 2$ and penalty values one-half of those shown in Table 5.5 produce results similar to those collected in Table 5.5.

5.6.2   Spherical k-means

An application of Algorithm 7 with zero penalty and `tol` $= 0.001$ to the PDDP generated partition does not change the confusion matrix given by Table 5.1. The decrease of penalty to $p = -0.1$ slightly improves the confusion matrix (see Table 5.6). With penalty $p = -0.4$ the algorithm generates the perfect diagonal confusion matrix (see Table 5.4). Further decrease in penalty does not change

Table 5.2   PDDP followed by Algorithm 4 with $p = 0$ generated 'confusion' matrix with **75** 'misclassified' documents.

Cluster/DocCol	DC0	DC1	DC2
Cluster 0	1437	22	9
Cluster 1	1	1360	5
Cluster 2	22	16	1019

Table 5.3   PDDP followed by Algorithm 4 with $p = 0.01$ generated 'confusion' matrix with **40** 'misclassified' documents.

Cluster/DocCol	DC0	DC1	DC2
Cluster 0	1453	17	8
Cluster 1	1	1377	4
Cluster 2	6	4	1021

Table 5.4   PDDP followed by Algorithm 4 with $p = 0.09$ generated 'confusion' matrix with **0** 'misclassified' documents.

Cluster/DocCol	DC0	DC1	DC2
Cluster 0	1460	0	0
Cluster 1	0	1398	0
Cluster 2	0	0	1033

Table 5.5   Penalty vs. 'misclassification' with $r_0 = r_1 = r_2 = 1$.

Penalty	Misclassification
0.00	75
0.01	40
0.02	20
0.03	17
0.04	8
0.05	5
0.06	4
0.07	2
0.08	1
0.09	0

Table 5.6   PDDP followed by Algorithm 7 with $p = -0.1$ generated 'confusion' matrix with **228** 'misclassified' documents.

Cluster/DocCol	DC0	DC1	DC2
Cluster 0	1375	13	6
Cluster 1	6	1376	115
Cluster 2	2	79	912

Table 5.7   Penalty vs. 'misclassification' with $r_0 = r_1 = r_2 = 1$.

Penalty	Misclassification
0.0	250
−0.1	228
−0.2	59
−0.3	4
−0.4	0

this result. The values of penalty versus 'misclassification' of final partitions generated by Algorithm 7 with `tol` $= 0.001$ are given in Table 5.7. In these experiments $r_0 = r_1 = r_2 = 1$.

5.7   Conclusion

The chapter presents three clustering algorithms: constrained k-means, constrained spherical k-means, and constrained `smoka`. Each algorithm is capable of clustering a vector dataset equipped with must-link constraints and a penalty function that penalizes violations of cannot-link constraints.

Numerical experiments with the first two algorithms show improvement of clustering performance in the presence of constraints. At the same time a single iteration of each algorithm changes the cluster affiliation of one vector only. A straightforward application of the algorithms to large datasets is, therefore, impractical.

In contrast, a single iteration of the proposed constrained smoka clustering changes all k clusters. Numerical experiments with constrained smoka and large datasets with must-link and cannot-link constraints will be reported elsewhere. Judicious selection of constraints is of paramount importance to the success of clustering algorithms. We plan to perform and report experiments with large datasets equipped with cannot-link and must-link constraints in the near future.

References

Banerjee A, Merugu S, Dhillon IS and Ghosh J 2005 Clustering with Bregman divergences. *Journal of Machine Learning Research* **6**, 1705–1749.

Basu S, Banerjee A and Mooney R 2004 Active semi-supervision for pairwise constrained clustering. *Proceedings of SIAM International Conference on Data Mining*, pp. 333–344.

Basu S, Davidson I and Wagstaff K 2009 *Constrained Clustering*. Chapman & Hall/CRC.

Berry M and Browne M 1999 *Understanding Search Engines*. SIAM.

Boley DL 1998 Principal direction divisive partitioning. *Data Mining and Knowledge Discovery* **2**(4), 325–344.

Brucker P 1978 On the complexity of clustering problems. *Lecture Notes in Economics and Mathematical Systems, Volume 157* Springer pp. 45–54.

Dhillon IS and Modha DS 1999 Concept decompositions for large sparse text data using clustering. Technical Report RJ 10147, IBM Almaden Research Center.

Dhillon IS and Modha DS 2001 Concept decompositions for large sparse text data using clustering. *Machine Learning* **42**(1), 143–175. Also appears as IBM Research Report RJ 10147, July 1999.

Dhillon IS, Kogan J and Nicholas C 2003 Feature selection and document clustering. In *Survey of Text Mining* (ed. Berry M), pp. 73–100. Springer.

Duda RO, Hart PE and Stork DG 2000 *Pattern Classification* second edn. John Wiley & Sons, Inc.

Kogan J 2007a *Introduction to Clustering Large and High–Dimensional Data*. Cambridge University Press.

Kogan J 2007b Scalable clustering with smoka. *Proceedings of International Conference on Computing: Theory and Applications*, pp. 299–303. IEEE Computer Society Press.

Marroquin J and Girosi F 1993 Some extensions of the k-means algorithm for image segmentation and pattern classification. Technical Report A.I. Memo 1390, MIT, Cambridge, MA.

Nasraoui O and Krishnapuram R 1995 Crisp interpretations of fuzzy and possibilistic clustering algorithms. *Proceedings of 3rd European Congress on Intelligent Techniques and Soft Computing*, pp. 1312–1318, Aachen, Germany.

Rockafellar RT 1970 *Convex Analysis*. Princeton University Press.

Rose K, Gurewitz E and Fox C 1990 A deterministic annealing approach to clustering. *Pattern Recognition Letters* **11**(9), 589–594.

Teboulle M 2007 A unified continuous optimization framework for center-based clustering methods. *Journal of Machine Learning Research* **8**, 65–102.

Teboulle M and Kogan J 2005 Deterministic annealing and a k-means type smoothing optimization algorithm for data clustering. In *Proceedings of the Workshop on Clustering High Dimensional Data and its Applications (held in conjunction with the Fifth SIAM International Conference on Data Mining)* (ed. Dhillon I, Ghosh J and Kogan J), pp. 13–22. SIAM, Philadelphia, PA.

Teboulle M, Berkhin P, Dhillon I, Guan Y and Kogan J 2006 Clustering with entropy-like k-means algorithms. In *Grouping Multidimensional Data: Recent Advances in Clustering* (ed. Kogan J, Nicholas C and Teboulle M) Springer pp. 127–160.

Wagstaff K and Cardie C 2000 Clustering with instance-level constraints. *Proceedings of the Seventeenth International Conference on Machine Learning*, pp. 1103–1110, Stanford, CA.

Wagstaff K, Cardie C, Rogers S and Schroedl S 2001 Constrained k-means clustering with background knowledge. *Proceedings of the Eighteenth International Conference on Machine Learning*, pp. 577–584, San Francisco, CA.

Zhang T, Ramakrishnan R and Livny M 1997 BIRCH: A new data clustering algorithm and its applications. *Journal of Data Mining and Knowledge Discovery* **1**(2), 141–182.

Part II

ANOMALY AND TREND DETECTION

6

Survey of text visualization techniques

Andrey A. Puretskiy, Gregory L. Shutt and Michael W. Berry

6.1 Visualization in text analysis

Visualization has been proven to be a very powerful tool in a wide variety of fields, including text mining. While text mining can reduce an enormous quantity of data to a significantly smaller subset, this subset is often still much too large for a human analyst to reasonably process, comprehend, detect trends, and draw conclusions from. Text visualization and visual text mining postprocessing tools can therefore be of crucial importance in facilitating knowledge discovery, as well as providing a *big picture* overview of overwhelmingly large amounts of data. This chapter explores several such visual techniques and describes specific examples of software that utilizes them.

There exist many different purposes for text visualization, dependent upon the user's needs at a particular time. One major purpose of visualization is to facilitate the tracing of alterations performed upon a document or set of documents over time. This may focus upon changes to the content of the document(s), or on authorship tracking. Visualizations in this category typically use variations of the time line plot technique, which typically involves constructing a color-coded plot that traces the changes made by each individual author over time. In applications

Text Mining: Applications and Theory   edited by Michael W. Berry and Jacob Kogan
© 2010, John Wiley & Sons, Ltd

where many different authors may be collaborating on a single document, this often results in an incredibly complex and difficult to read plot.

Sometimes a quick, complete, and graphical summary of a large document is all that the user requires. Tag clouds and other similar techniques have proven highly useful in this area. A tag cloud is a summary of a document or a collection of documents that relies upon font size, color, and/or text placement to indicate the relative importance of key terms to the user. The key terms may be chosen according to any number of schemes, some as simple as a straightforward term count. Though perhaps not particularly useful for detailed analysis, a tag cloud is highly effective in summarizing large amounts of text in an easily readable, and understandable, visual manner.

Another major purpose of text visualization is general text exploration: that is, a general search for interesting patterns or relationships within the data. Quite often, the user has very limited prior information regarding the target of his or her search, thus the term 'exploration' describes this type of analysis better. In order to facilitate it, visualization software in this category typically creates an altered, graphical term space representation – for example, an interconnected graph of all of the terms in a book, where terms may be connected based on co-occurrence within a single chapter or section. Many variations of this approach exist, but one aspect that most of them have in common is that they are heavily reliant upon the user's attention and perception. The user's ability to notice, interpret, and understand patterns in the dataset is a critical part of the analysis process when such software is utilized.

Sentiment tracking (and its related visualization software) is a relative new-comer to the text visualization arena, and yet it is a highly promising technique that has a great capability for insightful analysis of textual data. Various techniques for sentiment tracking exist. One common approach attempts to connect adjectives from the text to one of a number of basic emotion descriptor adjectives via a thesaurus synonym path. The length of the connecting paths determines how each text adjective is categorized. A percentage breakdown plot may then be constructed to indicate the overall content of basic emotions or sentiments within the text over time.

Many text mining procedures produce unlabeled, textual results (e.g. groups of interrelated terms that describe features contained in the original input dataset). In order to draw potentially useful conclusions, further interpretation of these results is necessary. This often requires a great commitment of time and effort on the part of human analysts. Visual postprocessing tools tailored for specific text mining packages can therefore greatly facilitate the analysis process. This chapter will discuss one such visual tool, *FutureLens*, in great detail.

6.2 Tag clouds

Conceptually, tag clouds are somewhat similar to histograms; however, they offer greater flexibility for the visual representation of the relative importance of each activities analytics article berry chinchilla collection contest data dataset discovery documents entities example factorization featurelens figure futurelens gil groups http identify information key mining nario news nonnegative ntf output patterns phrase plot prototype quickly relevant sce- scenario section shown software tensor terms text tool tracking university used user vast visual

Figure 6.1   A tag cloud of the paper in Shutt et al. (2009), generated by the TagCrowd application.

tag. The font size and color, as well as the orientation of the text (vertical or horizontal) and the proximity of tags to one another, may be used to convey information to the observer (Kaser and Lemire 2007). A basic tag cloud generator is a relatively simple and straightforward program that obtains term counts from textual data, then generates HTML that takes the term counts into consideration. Frequently, the user is allowed to choose the total number of terms in the tag cloud summary. The tag cloud generating code then selects these terms based on the overall counts and generates HTML code where font sizes vary according to the relative relationship between the overall term counts. Figure 6.1 demonstrates a straightforward and easy-to-use tag cloud generator application, TagCrowd (Steinbock 2009). The text of the paper in Shutt et al. (2009) was used to generate the tag cloud in the figure.

Figures 6.2 and 6.3 demonstrate a more complex application, Wordle (Feinberg 2009). This generator includes many additional graphical capabilities. It

Figure 6.2   A tag cloud of the paper in Shutt et al. (2009), generated by the Wordle application using the 'Vigo' font type and a randomized predominant text orientation.

Figure 6.3 A tag cloud of the paper in Shutt et al. (2009), generated by the Wordle application using the 'Boope' font type and with the predominant text orientation set to horizontal.

gives the user the ability to alter text and background color in a variety of ways. Font type may be modified. The predominant orientation of the words in the word cloud may be set in a variety of ways, ranging from completely horizontal, to mostly horizontal or mostly vertical, to completely vertical. Wordle is capable of automatically randomizing all of these parameters.

Both Steinbock (2009) and Feinberg (2009), as well as many other tag cloud generators, allow free noncommercial use of the images and/or HTML code that they generate. TagCrowd and Wordle both use the Creative Commons license, meaning users are allowed to copy, distribute, and transmit the materials (Commons 2009a,b). While Wordle does not limit usage to noncommercial applications, TagCrowd allows noncommercial use only. It should be noted that the source code of the generators is copyrighted by the respective authors and does not fall under the Creative Commons license.

6.3 Authorship and change tracking

The development of authorship tracking visual software was motivated by Wikipedia-like collaborative environments, where multiple users may make incremental changes to a single document over a relatively long time period. Software such as History Flow, a project of the Collaborative User Experience Research Group at IBM, allows the user to visually trace the changes to a particular document. The software creates a series of color-coded bars (by author), each corresponding to a single version or revision of the document. Same-color segments on adjacent bars are connected, creating a three-dimensional visual effect that provides the user with information on the way the document was altered over time by multiple authors. History Flow also includes additional visualization modes that allow the user to track a single author's activity through the collaboratively developed document, as well as to trace the changes by their relative age. IBM researchers have used History Flow to effectively study cooperation and conflict among authors on Wikipedia, including such aspects as vandalism and repair (Viégas et al. 2004). More information on History Flow, including screenshots of the software in action, may be found at Viégas et al. (2009).

6.4   Data exploration and the search for novel patterns

TextArc uses JavaScript and functions as an online application to visualize complex textual datasets. It has been applied to works of literature, such as *Alice in Wonderland* and *Hamlet*. The visualization provided by TextArc consists of two levels. First, the original text is available around the periphery of the visualization area. Second, an interconnected graph of terms is provided in the middle of the visualization area. The two areas are interconnected, meaning that the user is able to select any particular term in the middle area and quickly see its context in the full text that is displayed along the periphery. This software allows the user to easily determine any given term's relevance or relative importance to any part of the literary work (Paley 2009). Figures 6.4 and 6.5 demonstrate how TextArc was used to explore Shakespeare's *Hamlet*.

6.5   Sentiment tracking

Sentiment tracking involves tracing an author's changing attitudes through a particular piece of text. In order to accomplish this, it is necessary to categorize the terms from the text to certain broad descriptor adjectives. Descriptor adjectives may vary: for example, the SEASR (Software Environment for the Advancement

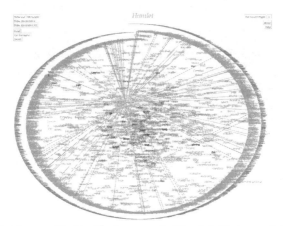

Figure 6.4   TextArc applied to Shakespeare's Hamlet. *Not surprisingly, the name 'Hamlet' figures prominently in the work.*

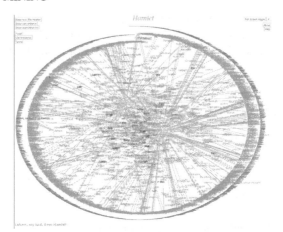

Figure 6.5   TextArc allows the user to easily track the connections between various terms. Here, we see that the term 'Hamlet' is related to the term 'lord'. It is also possible to track either term further.

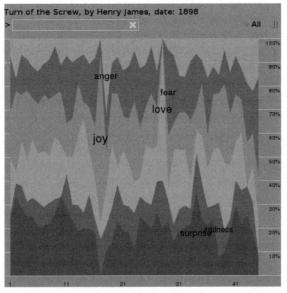

Figure 6.6   SEASR's Sentiment Tracking project applied to Turn of the Screw, *by Henry James (1898). Each unit on the X-axis corresponds to a group of 12 sentences. The Y-axis shows the sentiment composition for all six of Parrott's core emotions (Parrott 2000).*

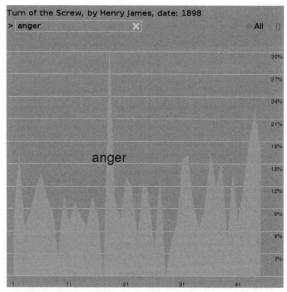

Figure 6.7   SEASR's Sentiment Tracking project applied to Turn of the Screw, *by Henry James (1898). This figure shows the presence of anger in the literary work.*

of Scholarly Research) Sentiment Tracking project used Parrot's six core emotions in its sentiment tracking demonstration (Figures 6.6, 6.7, and 6.8): Love, Joy, Surprise, Anger, Sadness, and Fear (Parrott 2000). The Sentiment Tracking project uses UIMA (Unstructured Information Management Applications), a component framework for analyzing unstructured content, including but not limited to text. UIMA began as a project at IBM, but evolved into an open source project at the Apache Software Foundation (SEASR 2009b). Several different metrics may be used in order to categorize the terms from the text. The approach used by the SEASR/UIMA Sentiment Tracking project involves searching for the shortest path through a thesaurus from each term within the text to one of the descriptor adjectives. Synonym symmetry is another useful technique, and may be helpful as a 'tie breaker' (SEASR 2009a).

6.6   Visual analytics and FutureLens

FutureLens is a Java-based visual analytics environment that has been used to support the extraction and tracking of scenarios and plots from news articles defining the VAST 2007 Contest (Scholtz et al. 2007). Using groups of related persons, locations, organizations, and context-specific words and phrases identified (through time) by nonnegative tensor factorization (NTF) models (Bader et al. 2008b), FutureLens was instrumental in extracting the

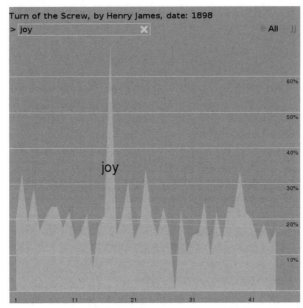

Figure 6.8   SEASR's Sentiment Tracking project applied to Turn of the Screw, *by Henry James (1898). This figure shows the presence of joy in the literary work.*

underlying (fictitious) criminal and terrorist activities created by Whiting et al. for the VAST 2007 Contest. Section 6.7 briefly describes the scenario mining process and expectations that warrant the design of visual analytic software like FutureLens. An early prototype of FutureLens is discussed in Section 6.8, followed by an illustration of some of the important features of FutureLens in Section 6.9. Examples of scenario discovery with the VAST 2007 Contest dataset are provided in Section 6.10 and Section 6.11. A brief discussion of future enhancements to FutureLens is given in Section 6.12 (Shutt et al. 2009).

6.7   Scenario discovery

The intent of the IEEE VAST 2007 Contest (Scholtz et al. 2007) was to promote the development of benchmark datasets and metrics for visual analytics as well as to establish a forum for evaluating different solution strategies. In providing news stories, blog entries, background information, and limited multimedia materials (small maps and data tables), the contest organizers challenged the participants to investigate a major law enforcement/counter-terrorism scenario, form a hypothesis, and collect supporting evidence. Tasks that each team/entry was expected to address included: (1) identify entities (e.g. people, places, and activities) from text and multimedia information; (2) develop interactive tools to visualize/analyze this information; (3) answer specific (contest-provided) questions based on the analysis; and (4) produce a video that demonstrates how those answers were derived. FutureLens was primarily used for the second task to visualize and track the entity groups generated by the nonnegative tensor factorization models discussed in Bader et al. (2008a,b).

6.7.1 Scenarios

The primary (crime and terrorism-based) scenarios depicted in the VAST 2007 Contest involved wildlife law enforcement incidents occurring in the fall of 2004. Endangered species issues and ecoterrorism activities played key roles in the underlying terrorist scenario/plot. The data used to describe the details of the plot included text, images, and some statistics. Although activities of certain animal rights groups, such as the People for the Ethical Treatments of Animals (PETA) and Earth Liberation Front (ELF), were involved with the plot, the contest organizers did not consider them to be the primary (interesting) parties for investigation. In fact, such sideplots were used to deflect attention from the main criminal/terrorist scenarios, thus providing a realistic challenge.

6.7.2 Evaluating solutions

Although entries (or answers) submitted to the VAST 2007 Contest were judged according to the correctness of the answers to the questions and the evidence provided, a more subjective assessment of the quality of the displays, interactions, and support for the analytical process was also provided. The last category is of particular interest because the field of text mining, in general, could greatly benefit from the design of more intuitive visualizations that expose or verify potential scenarios of human activity.

Following the traditional cues of journalistic reporting, visual analytics (as reflected by the VAST 2007 contest) seeks to answer the questions (who, what, where, and when) for an alleged activity using the the most relevant documents or other materials from the dataset as evidence. Contest participants were required to describe the plot(s) and subplots(s) and how people, motivations, activities, and locations relate to the plot; that is, their relationships, and any uncertainties or information gaps that exist. For example, some of the questions each entry was required to answer include:

- **(Who)** Who are the players engaging in questionable activities in the plot(s)? When appropriate, specify the organization they are associated with.

- **(When/What)** What events occurred during this time frame that are most relevant to the plot(s)?

- **(Where)** What locations are most relevant to the plot(s)?

6.8   Earlier prototype

Many of the concepts and ideas of this project stem from FeatureLens, a University of Maryland (Human–Computer Interaction Laboratory) text and pattern visualization program (Don et al. 2007, 2008; Kumar 2009). FeatureLens allows the user to explore frequently occurring terms or patterns in a collection of documents. Connections between these frequent terms and the dates at which they appear in the set of documents can quickly be visualized and investigated. A screenshot of the FeatureLens prototype is shown in Figure 6.9.

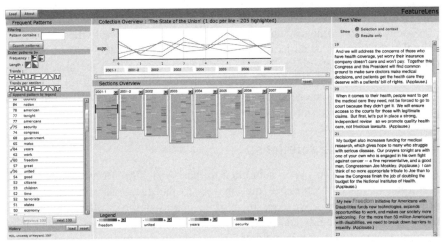

Figure 6.9   FeatureLens prototype (written in Ruby) developed at the University of Maryland Human–Computer Interaction Laboratory.

While FeatureLens may sound suitable for the given task, it is not without its shortcomings. For one, its design is rather complex as it requires a MySQL database server, an HTTP server, and an Adobe Flash-enabled web browser to function properly. As such, it is not a trivial task to set up an instance of FeatureLens from scratch and may take an inexperienced user a significant amount of time to get started. Datasets must be parsed and stored in the database, an operation that an end user cannot perform, so examining arbitrary datasets is out of the question. In implementing the architecture of FeatureLens, the designers chose to use a variety of languages: Ruby for the back end, XML to communicate between the front end and back end, and OpenLaszlo for the interface. Because of this variety in languages, adapting and modifying FeatureLens would prove quite difficult. Responsiveness of the interface also tends to degrade to the point that it impacts usability when given even the simplest of tasks. Clearly a better solution was needed.

6.9   Features of FutureLens

FutureLens is a text visualization tool that implements much of the functionality of FeatureLens while adding several necessary features. The most significant among the additional features is the capability to create term collections and phrases. The user may do this by simply clicking and dragging selected terms or entities onto each other. FutureLens is written in the Java programming language using the Standard Widget Toolkit so it is not only cross-platform but uses native widgets where possible to maintain a look and feel consistent with the users' platform. For end users not familiar with the program, FutureLens has a built-in feature that demonstrates its basic functionality. An example of FutureLens running under Mac OS X is shown in Figure 6.10.

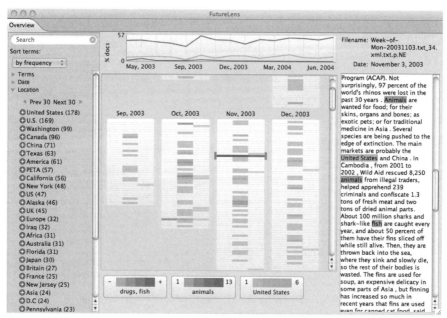

Figure 6.10   FutureLens prototype (written in Java) developed at the University of Tennessee for visualization of NTF-generated outputs.

All the basic functionality of FutureLens can be seen in this example. The boxes along the bottom show the terms that are currently being investigated. The intensity of the color in these boxes hints at the concentration of the term throughout the documents. A graph of the percentage of documents containing the term versus time is shown at the top, while the raw text of the selected document is shown to the right with the selected terms highlighted in the appropriate color.

Multiple terms can easily be combined into extended patterns by dragging and dropping. Terms may be combined into either collections or phrases. A collection is created when the user drags and drops terms onto each other. Term adjacency does not affect search results for a collection. If the users holds down the COPY key (this key varies depending on the operating system; for example, on Mac OS X this is the ALT key), a phrase rather than a collection will be created. In this case, term adjacency will be considered when the software performs searching. While this presents an excellent overview of the data, it is also possible to load the output (groups of terms and/or entities) derived from a data clustering method. An example of this is shown in Figure 6.11.

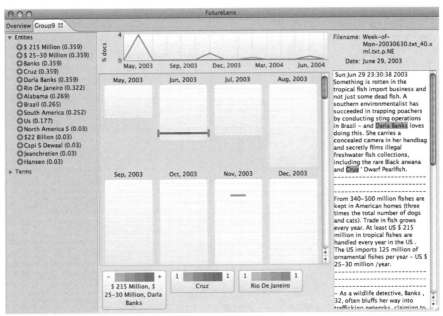

Figure 6.11   FutureLens tracking the co-occurrences of grouped terms and entities (persons, locations, and organizations).

Here a file containing pertinent terms output from a nonnegative tensor factorization (NTF) tool has been loaded as a separate view into FutureLens. The view is nearly identical to the overview. However, the list of terms has been limited to only what was contained in the input file. This allows the user to quickly view the different clusters of entities through time (Bader et al. 2008b).

6.10   Scenario discovery example: bioterrorism

Figures 6.12 through 6.16 demonstrate how FutureLens may be used together with NTF to quickly reconstruct a bioterrorism-related plotline that was buried within the VAST 2007 text corpus. In Figure 6.12, one of the NTF output groups has been loaded into FutureLens. Each NTF output group contained 15 top ranking (most relevant) entities and 35 top ranking terms that described a particular feature of the input dataset. The user is aware that he or she should be searching for some sort of interesting and nefarious scenario. The selected terms (*Monkeypox*, *Exotic*, *Pets*, *Chinchilla*) constitute a good starting point. However, the user will not find all news articles with the occurrence of the relatively common words *Pets* and *Exotic* relevant. Thus, the two terms are combined into the phrase *Exotic Pets*, as shown in Figure 6.13. Figure 6.14 demonstrates how FutureLens allows

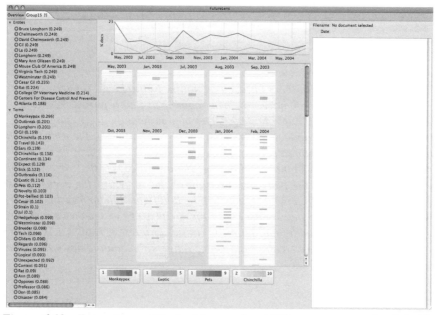

Figure 6.12   FutureLens with the bioterrorism NTF output group loaded. The panel on the left shows the terms and entities relevant to the NTF output group. The top-level graph summarizes the frequency of the selected terms and entities over time. The monthly frequency plots in the center of the screen allow the user a more detailed view of the term/entity occurrence over time. The monthly plots are clickable; the results of that operation are demonstrated in the subsequent figures.

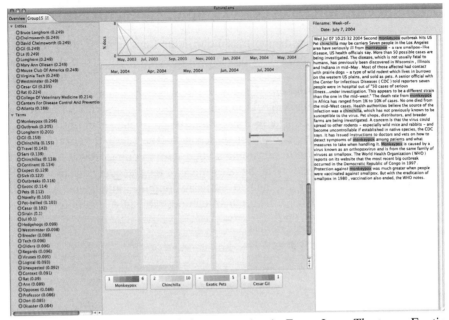

Figure 6.13   Demonstration of phrase creation in FutureLens. The terms Exotic *and* Pets *from Figure 6.12 have now been combined into a phrase* Exotic Pets. *The phrase creation technique has the effect of significantly decreasing the total number of hits, thereby reducing on-screen clutter and allowing the user to focus his or her search. Additionally, this figure demonstrates the effect of the user's clicking one of the bars in the monthly plots. Doing so causes the corresponding text to be displayed in the panel on the right of the screen. If the user-selected terms are contained within the text, they will be highlighted in appropriate colors. This allows the user to quickly ascertain the context of the selected terms, and possibly also to locate additional terms or entities of interest. A phrase is created when the user drags selected terms onto each other while holding down the* COPY *key (e.g.* ALT *on Mac OS X).*

the user to easily identify a key news story within the large dataset. The article shown in this figure contains a great amount of relevant information regarding an outbreak of a potentially deadly virus, monkeypox, in the Los Angeles area. The article implies that the outbreak may not have been accidental, and connects it to an animal rights activist and chinchilla breeder named Cesar Gil. In order to fully reconstruct the plotline, the user selects the names *Cesar Gil* and *Gil* from the Entities list, as shown in Figure 6.15. However, this results in too many instances of *Gil* being found, and most of them are probably irrelevant. Exploiting the link between Gil and chinchilla breeding, the user combines the terms *Chinchilla* and *Gil* into a collection. This helps the user to quickly identify a relevant article that contains an advertisement for Gil's chinchilla breeding business (Figure 6.16).

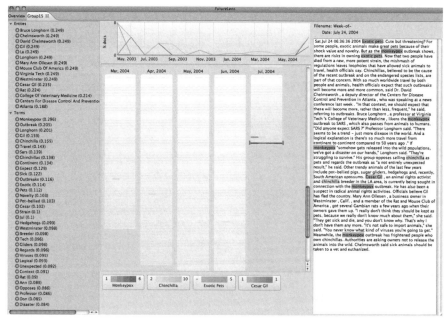

Figure 6.14   Key news story identification using FutureLens. The monthly plots allow for convenient visualization of term co-occurrence over time. As demonstrated in this figure, term co-occurrence allows the user to quickly extract the most relevant and informative textual data from a large dataset. In this example, the news article that contains all of the user's selected terms contains a great deal of information relevant to the chinchilla–bioterrorism plot. The context provided by the article tells the user exactly in what way many of the terms and entities within the NTF output group are relevant to the bioterrorism scenario that was hidden within this textual dataset.

Not all of the articles that are relevant to this plotline have been shown in the figures; however, FutureLens enables the user to quickly and easily identify them all. FutureLens also helps the user to focus on the relevant parts of the article (Shutt et al. 2009).

6.11   Scenario discovery example: drug trafficking

Figures 6.17 through 6.21 demonstrate how FutureLens may be used together with NTF to quickly reconstruct a drug trafficking plotline that was buried within the VAST 2007 text corpus. In Figure 6.17, the corresponding NTF output group has been loaded into FutureLens. Figure 6.18 shows both term chaining techniques: *Tropical* and *Fish* are combined into the phrase *Tropical Fish*; *Cocaine* and *Drugs* are combined into a single collection of terms. As a result of this operation several news articles are found, including one that

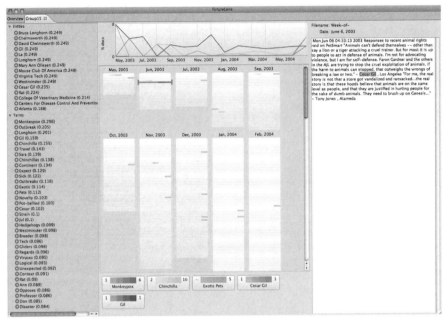

Figure 6.15   Entity of interest search using FutureLens. The key news article demonstrated in Figure 6.14 revealed that an individual named Cesar Gil *is a key player in this scenario. FutureLens allows the user to expand the search by including alternative forms of this individual's name (e.g.* Gil). *However, this may cause a significant number of irrelevant search results. Figure 6.16 demonstrates how the user might use FutureLens' collection creation capability to focus the search.*

discusses the use of trade in exotic pets (including tropical fish) as a cover for drug smuggling (including cocaine trafficking). The next figure, Figure 6.19, shows the selection of what appears to be a company name, *Global Ways*, from the Entities list. As shown, the user is able to quickly find a story that identifies *Global Ways* as a company that imports exotic tropical fish from South America into the United States. Given the previously established connection between drug trafficking and tropical fish imports, Global Ways may be worth investigating further. As Figure 6.20 shows, shortly after publication of the story advertising Global Ways' import business, the Fish and Wildlife Service had issued a warning to avoid handling shipments of tropical fish that may have entered the United States through Miami. According to this story, the packaging of some of these shipments appears to have been contaminated with an unknown toxic substance. Global Ways is listed as one of the suspects.

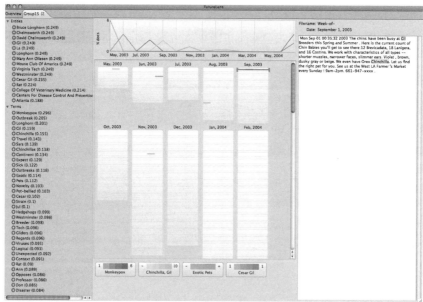

Figure 6.16   Term collection creation in FutureLens. A collection of terms may be created in FutureLens by simply dragging selected terms onto each other. A collection of terms differs from a phrase because term adjacency does not matter for a collection search. In this example, the user has been able to determine that the Gil *of interest is highly likely to be mentioned in a news article that also contains the term* chinchilla. *The user created a collection containing both terms, thereby greatly reducing the total number of search hits on the term* Gil *alone. Furthermore, this leads to the discovery of a highly relevant article, one in which the individual named* Gil *is advertising the sale of chinchillas that would later be proved to have been intentionally infected with the potentially deadly monkeypox virus.*

Finally, Figure 6.21 identifies the owner of Global Ways as *Madhi Kim*, thereby allowing the analyst to continue tracing relationships through the dataset.

6.12   Future work

While FutureLens provides numerous features for plot and scenario discovery, there is still room for improvement. It works well for evidence generation but it has no automation for any type of scenario discovery. Methods that locate

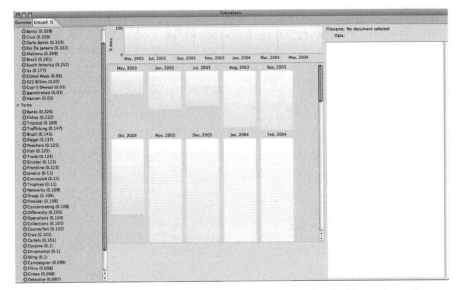

Figure 6.17   The drug trafficking NTF output group loaded into FutureLens.

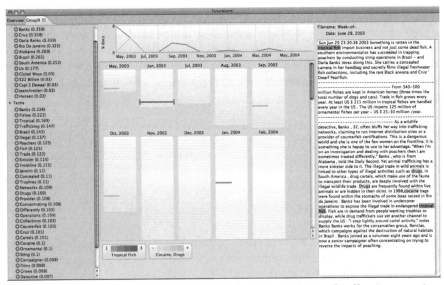

Figure 6.18   Two types of term chaining, phrase creation and collection creation, help the user to quickly identify relevant news stories.

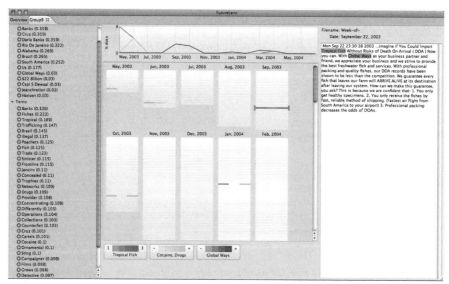

Figure 6.19   Among entities of interest produced by NTF, there appears a com-pany name, Global Ways. *FutureLens enables the user to further explore the relationship between this company, the tropical fish trade, and drug trafficking.*

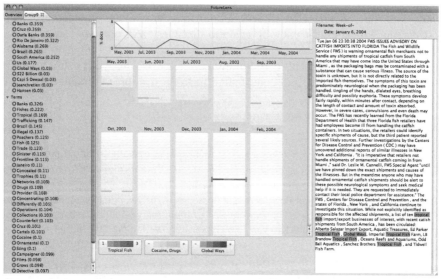

Figure 6.20   FutureLens helps the user to identify news stories that connect Global Ways *to drug trafficking.*

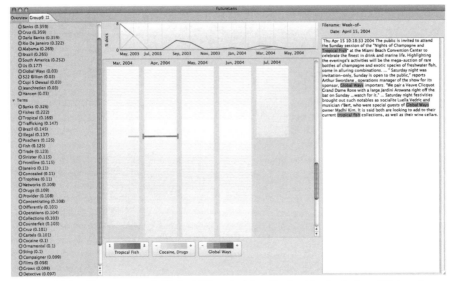

Figure 6.21   The owner of Global Ways is identified with the help of FutureLens. Further investigation of the owner's connections and associations is possible at this point.

interesting features in the dataset could be added to create a single analysis tool. As it stands now, the output of data mining models such as that created by nonnegative tensor factorization (see Bader et al. (2008b)) must be entered manually into the software environment. Eliminating this human interaction would greatly increase the efficiency of scenario discovery. An obvious extension for dynamic (time-varying) datasets is certainly needed. The portability and intuitive word/phrase tracking capability of FutureLens, however, make this public-domain software environment a solid contribution to the text mining community.

References

Bader BW, Berry MW and Browne M 2008a Discussion tracking in Enron email using PARAFAC. In *Survey of Text Mining II: Clustering, Classification, and Retrieval* (ed. Berry M and Castellanos M) Springer-Verlag pp. 147–163.

Bader BW, Puretskiy AA and Berry MW 2008b Scenario discovery using nonnegative tensor factorization. In *Progress in Pattern Recognition, Image Analysis and Applications* (ed. Ruiz-Shulcloper J and Kropatsch WG) Springer-Verlag pp. 791–805.

Commons C 2009a Creative Commons Attribution License 3.0 http://creativecommons.org/licenses/by/3.0/us/.

Commons C 2009b Creative Commons Non-Commercial Attribution license 3.0 http://creativecommons.org/licenses/by-nc/3.0/.

Don A, Zhelev E, Gregory M, Tarkan S, Auvil L, Clement T, Shneiderman B and Plaisant C 2007 Discovering interesting usage patterns in text collections: integrating text mining with visualization. *HCIL Technical Report 2007-08*.

Don A, Zheleva E, Gregory M, Tarkan S, Auvil L, Clement T, Shneiderman B and Plaisant C 2008 Exploring and visualizing frequent patterns in text collections with FeatureLens. http://www.cs.umd.edu/hcil/textvis/featurelens. Visited November 2008.

Feinberg J 2009 Wordle: Beautiful word clouds. http://www.wordle.net. Visited July 2009.

Kaser O and Lemire D 2007 Tag-cloud drawing: Algorithms for cloud visualization. *CoRR*.

Kumar A 2009 The MONK Project Wiki. https://apps.lis.uiuc.edu/wiki/display/MONK/The+MONK+Project+Wiki. Last edited August 2008.

Paley WB 2009 TextArc. http://www.textarc.org/. Visited July 2009.

Parrott WG 2000 Emotions in social psychology: Volume overview. In *Emotions in Social Psychology: Essential readings* (ed. Parrott WG) Psychology Press pp. 1–19.

Scholtz J, Plaisant C and Grinstein G 2007 IEEE VAST 2007 Contest. http://www.cs.umd.edu/hcil/VASTcontest07.

SEASR 2009a Sentiment tracking from UIMA data. http://seasr.org/documentation/uima-and-seasr/sentiment-tracking-from-uima-data/. Visited July 2009.

SEASR 2009b UIMA and SEASR. http://seasr.org/documentation/uima-and-seasr/. Visited July 2009.

Shutt GL, Puretskiy AA and Berry MW 2009 FutureLens: Software for text visualization and tracking *Text Mining Workshop, Proceedings of the Ninth SIAM International Conference on Data Mining, Sparks, NV*.

Steinbock D 2009 TagCrowd: Create your own tag cloud from any text to visualize word frequency. http://www.tagcrowd.com. Visited July 2009.

Viégas FB, Wattenberg M and Dave K 2004 Studying cooperation and conflict between authors with History Flow visualizations *Proceedings of the SIGCHI Conference on Human Factors in Computing Systems*, pp. 575–582. ACM Press.

Viégas FB, Wattenberg M and Dave K 2009 History Flow: Visualizing the editing history of Wikipedia pages http://www.research.ibm.com/visual/projects/history_flow/index.htm.

7

Adaptive threshold setting for novelty mining

Wenyin Tang and Flora S. Tsai

7.1 Introduction

In the age of information, it is easy to accumulate various documents such as news articles, scientific papers, blogs, advertisements, etc. These documents contain rich information as well as useless or redundant information. People who are interested in a certain topic may only want to track the new developments of an event or the different opinions on the topic. This motivates the study of novelty mining, or novelty detection, which aims to retrieve novel, yet relevant, information, given a specific topic defined by a user (Zhang and Tsai 2009a). A typical novelty mining system consists of two modules: (1) categorization; and (2) novelty mining. The categorization module classifies each incoming document into its relevant topic bin. Then, the novelty mining module detects the documents containing enough novel information in the topic bin. This chapter will focus on the later module. Due to its importance in information retrieval, a great deal of attention has been given to novelty mining in the past few years. The pioneering work for novelty mining was performed at the document level (Zhang et al. 2002). Later, more contributions were made to novel sentence mining, such as those reported in TREC 2002–2004 Novelty Track (Harman 2002; Soboroff 2004; Soboroff and Harman 2003), those in comparing various novelty metrics

Text Mining: Applications and Theory   edited by Michael W. Berry and Jacob Kogan
© 2010, John Wiley & Sons, Ltd

(Allan et al. 2003; Tang and Tsai 2009: Zhao et al. 2006), and those in integrating various natural language processing (NLP) techniques (Kwee et al. 2009; Ng et al. 2007; Zhang and Tsai 2009b).

Novelty mining is a process of mining novel text in the relevant documents of a given topic. The novelty of any document or sentence is quantitatively measured by a novelty metric based on its history documents and represented by a novelty score. The final decision on whether a document or sentence is novel or not depends on whether the novelty score falls above or below a threshold. As an adaptive filtering algorithm, novelty mining is one of the most challenging problems in information retrieval. One primary challenge is how to set the threshold of novelty scores adaptively. In the novelty mining system, since there is little or no training information available, the threshold cannot be predefined with confidence. The motivations for designing an adaptive threshold setting for the novelty mining system are manifold. There is little training information in the initial stages of novelty mining and different users may have different definitions about novelty. Motivations of adaptive threshold setting will be analyzed in detail later (in Section 7.2.2).

To the best of our knowledge, few studies have focused on adaptive threshold setting in novelty mining. A simple threshold setting algorithm was proposed in Zhang et al. (2002), which decreases the redundancy threshold a little if a redundant document is retrieved as a novel one based on a user's feedback. Clearly it is a weak algorithm because it can only decrease the redundancy threshold. This chapter addresses the problem of setting an adaptive threshold by modeling the score distributions of both novel and nonnovel documents. Although score distribution-based threshold-setting algorithms have been proposed for relevant document/sentence retrieval (Arampatzis et al. 2000, Robertson 2002; Zhai et al. 1999; Zhang and Callan 2001), the novelty score in novelty mining has its distinctive characteristics. In our experimental study, we find that scores from the novel and nonnovel classes heavily overlap. This is intuitive because novel and nonnovel information are always interlaced in one document, while in the relevance retrieval problem most of the nonrelevant documents show little similarity with relevant ones. Second, we find that the score distributions for both novel and nonnovel classes can be approximated by Gaussian distributions (detailed in Section 7.2.3). In the relevance retrieval problem, however, the scores of nonrelevant documents follow an exponential distribution (Arampatzis et al. 2000). This also implies that most nonrelevant documents are dissimilar to relevant ones.

The score distributions of classes provide the global information necessary for constructing an optimization criterion for threshold setting, while the threshold that optimizes this criterion is the best we can obtain until new user feedback is provided. Our proposed method, the Gaussian-based adaptive threshold setting (GATS) algorithm, is a general algorithm, which can be tuned according to different performance requirements, by employing different optimization criteria, such as the F_β score (Equation (7.7)), which is the weighted harmonic average of precision and recall where β controls the trade-off between them.

The novelty mining system combined with GATS has been tested on both document-level and sentence-level data and compared to the novelty mining system using various fixed thresholds. The experimental results show that a good performance of GATS can be obtained at both levels.

The remainder of this chapter is organized as follows. Section 7.2 first analyzes the motivations of threshold setting in novelty mining, and then introduces the GATS algorithm. Section 7.3 tests GATS at both the sentence level and document level. Section 7.4 concludes the chapter.

7.2 Adaptive threshold setting in novelty mining

7.2.1 Background

Novelty mining is the process of mining novel text in the relevant documents of a given topic. The novelty of a document or sentence (later we refer only to documents without losing any generalization) can be quantitatively measured by a novelty metric and represented by a novelty score. The most commonly used novelty metric, the cosine distance metric, will be employed throughout this chapter as it yielded good results for novelty mining compared to more complex metrics (Zhang et al. 2002). Since cosine similarity does not measure the degree of novelty directly, we convert the cosine similarity scores to novelty scores by subtracting these similarity scores from one. The cosine similarity novelty metric compares the current document to each of its history documents separately, whereas the minimum novelty score among them will be used as the novelty score of the current document. Specifically,

$$N_{\cos}(d_t) = \min_{1 \le i \le t-1} [1 - \cos(d_t, d_i)], \text{ where} \qquad (7.1)$$

$$\cos(d_t, d_i) = \frac{\sum_{k=1}^{n} w_k(d_t) \cdot w_k(d_i)}{\|d_t\| \cdot \|d_i\|},$$

and where $N_{\cos}(d)$ denotes the cosine similarity-based novelty score of document d and $w_k(d)$ is the weight of the kth word in document weighted vector d. The weighting function used in our work is the term frequency.

The final decision on whether a document is novel or not depends on whether the novelty score falls above or below a threshold. The document predicted as 'novel' will be pushed into the history document list.

When novelty mining adopts a fixed threshold, no user feedback is considered and the whole process is unsupervised. When novelty mining adopts an adaptive threshold setting algorithm, the system needs to respond to any new feedback from the user. Based on the feedback from the user, the new threshold output by this algorithm will replace the current one and be used for future incoming documents until new feedback is available. Note that when no feedback is received, the system will fix the threshold at the initial threshold.

7.2.2 Motivation

There are several reasons motivating us to design an adaptive threshold setting algorithm for novelty mining. First of all, there is little or no training information in the initial stages of novelty mining. Therefore, the threshold can hardly be predefined with confidence. The training information that is necessary for threshold setting includes the statistics of data and users' reading habits. For example, a topic with 90% novel documents needs a relatively low threshold for novelty scores to retrieve most of the documents. On the other hand, different users may have different definitions of 'novel' information. For example, one user might regard a document with 50% novel information as a novel document while another user might only regard a document with 80% novel information as a novel document. The threshold of novelty scores should be higher for the user with a stricter definition of the 'novel' document. As novelty mining is an accumulating system, more training information will be available for threshold setting, based on user feedback given over time. The adaptive threshold setting algorithm is able to utilize this available information and customizes the novelty mining system to the user's needs.

Satisfying different performance requirements is another important motivation for employing an adaptive threshold setting algorithm for novelty mining. For example, when users do not want to miss any novel information, a high-recall system that only filters out very redundant documents is desired. When users want to read the most novel documents first, a high-precision system that only retrieves very novel documents is preferred. Therefore, the threshold should be tuned according to different performance requirements.

Next, we will introduce the proposed method, namely GATS, and explain how it works with novelty mining.

7.2.3 Gaussian-based adaptive threshold setting

GATS is a score distribution-based threshold-setting method. It models the score distributions of both novel and nonnovel documents by Gaussian probability distributions. The score distributions of both classes provide global information on the data, from which we can construct an optimization criterion for searching the optimal threshold. Therefore, two major issues in GATS are: (1) modeling the novelty score distributions; and (2) constructing the optimization criterion for searching the best threshold. Next, we will introduce these two issues separately.

Novelty score distributions

Assume there are n training documents, $d_1, d_2, \ldots, d_n$, each of which belongs to either novel class c_1 or nonnovel class c_0. For any document d_i, $i = 1, 2, \ldots, n$, its novelty score, x_i, can be estimated by some novelty metric such as cosine similarity as defined in Equation (7.1).

To find the empirical novelty score distributions of data, some training datasets are needed. Here, we use topics N54 and N69 from the TREC 2004 Novelty Track data (Soboroff 2004) and assume all the novel and nonnovel documents are known beforehand. The following steps are processed on both training datasets separately.

Step 1: Calculate the novelty scores x_i of each document d_i, $i = 1, 2, \ldots, n$, using Equation (7.1), where the history document list includes all the history novel documents.

Step 2: Divide the scores of each class into several equal width bins with a bin width equal to [max(scores) − min(scores)]/no. of bins, where the number of bins for any class c_k equals the maximum integer smaller than $n_k/5$. Then, we can obtain the number of documents falling in the lth bin of the kth class, denoted as $n_{k,l}$, where, $l = 1, 2, \ldots$, no. of bins and $k \in \{c_0, c_1\}$.

Step 3: Obtain the empirical distributions of novelty scores, where the number of documents in each bin is normalized as follows:

$$p_e(x|c_k) = \frac{\text{no. of bins}}{n_k} \times n_{k,l}, \qquad (7.2)$$

where n_k and $n_{k,l}$ are the total number of documents in the class c_k and the number of documents falling in the lth bin of class c_k, respectively. The empirical distributions of novelty scores for topics N54 and N69 are shown in Figures 7.1 and 7.2, respectively.

The Gaussian distribution (also called the normal distribution) in the random variable X with mean μ and variance σ^2 has the probability density function (pdf)

$$p(x) = \frac{1}{\sigma\sqrt{2\pi}} e^{-\left(\frac{x-\mu}{\sigma\sqrt{2}}\right)^2}. \qquad (7.3)$$

If we assume that both novelty scores of the novel class and nonnovel class follow the Gaussian distributions, for any class c_k, $k \in \{0, 1\}$, the maximum likelihood estimations for the Gaussian probability density function $p(x|c_k) \sim G(\mu_k, \sigma_k^2)$ are given by

$$\mu_k = \frac{1}{n_k} \sum_{i \in c_k} x_i, \qquad (7.4)$$

$$\sigma_k^2 = \frac{1}{n_k} \sum_{i \in c_k} (x_i - \mu)^2. \qquad (7.5)$$

The Gaussian probability density function estimated for each class is represented by the dashed lines in Figures 7.1 and 7.2. It would appear that novelty

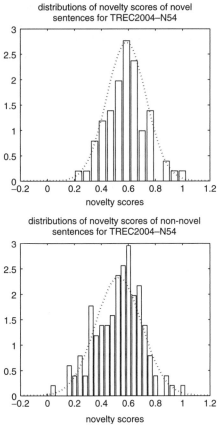

Figure 7.1   Empirical and probability distribution approximation for TREC 2004 Novelty Track data topic N54.

scores from both the novel and nonnovel classes can be well fitted by Gaussian distributions.

Optimization criterion

Assume we have an incoming document stream d_1, d_2, to d_n, of which n_1 are novel. After filtering the document stream by the novelty mining system with a threshold θ, any document can be classified in one of four classes as shown in Table 7.1.

Precision and recall are two widely used measures for evaluating the quality of results in information retrieval. Precision can be seen as a measure of exactness, whereas recall is a measure of completeness. In novelty mining, precision reflects how likely the system-retrieved documents are truly novel and recall reflects how likely the truly novel documents can be retrieved by the system. Precision and

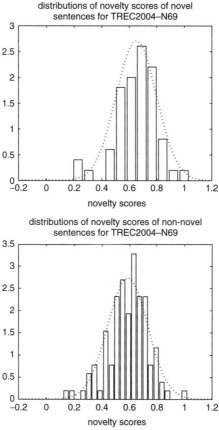

Figure 7.2   Empirical and probability distribution approximation for TREC 2004 Novelty Track data topic N69.

Table 7.1   Contingency table in the novelty mining system.

	Novel	Nonnovel
Retrieved	R_1	R_0
Nonretrieved	N_1	N_0
Total	n_1	n_0

recall for novel documents are defined as follows:

$$\text{precision} = \frac{R_1}{R_1 + R_0},$$

$$\text{recall} = \frac{R_1}{n_1}.$$

(7.6)

In novelty mining, the most commonly used evaluation measure is the F score (Soboroff 2004) (see also Section 3.4 in this book), which is the harmonic average of precision and recall:

$$F = \frac{2 \times \text{precision} \times \text{recall}}{\text{precision} + \text{recall}}. \qquad (7.7)$$

The F score is a special case of the F_β score, i.e. the weighted harmonic average of precision and recall

$$F_\beta = \frac{1}{\frac{\beta}{\text{precision}} + \frac{1-\beta}{\text{recall}}}, \qquad (7.8)$$

where β is the parameter to control the weights of precision and recall.

The numbers of documents in each class of the contingency table, R_1, R_0, N_1, and N_0, are functions of θ and can be approximated by the probability distributions of the novel and nonnovel classes. For example, given a threshold θ, the estimation of $R_1(\theta)$ is proportional to the probability of novel documents with novelty scores greater than the θ. Specifically,

$$R_1(\theta) = n_1 \cdot P(x > \theta | c_1) \qquad (7.9)$$

$$= n_1 \cdot \int_\theta^{+\infty} p(x|c_1)dx.$$

Similarly, we can obtain the other functions

$$R_0(\theta) = n_0 \cdot P(x > \theta | c_0), \qquad (7.10)$$

$$N_1(\theta) = n_1 \cdot P(x < \theta | c_1),$$

$$N_0(\theta) = n_0 \cdot P(x < \theta | c_0).$$

Substituting Equations (7.9) and (7.10) into Equation (7.6), precision and recall can be rewritten as functions of the threshold θ, as follows:

$$\text{precision}(\theta) = \frac{P_{c_1} P(x > \theta | c_1)}{P_{c_1} P(x > \theta | c_1) + P_{c_0} P(x > \theta | c_0)}, \qquad (7.11)$$

$$\text{recall}(\theta) = P(x > \theta | c_1), \qquad (7.12)$$

where P_{c_1} and P_{c_0} are the prior probabilities of the novel and nonnovel classes which can be estimated by

$$P_{c_1} = n_1/n \text{ and } P_{c_0} = n_0/n. \qquad (7.13)$$

After obtaining precision and recall as functions of θ, we can construct the optimization criterion for determining the best threshold. Substituting Equations

(7.11) and (7.12) into Equation (7.7), we can obtain the criterion $F_\beta(\theta)$, whose maximum value corresponds to the best threshold θ^*, i.e.

$$\theta^* = \arg \max_\theta F_\beta(\theta) \tag{7.14}$$

$$= \arg \max_\theta \frac{1}{\frac{\beta}{\text{precision}(\theta)} + \frac{1-\beta}{\text{recall}(\theta)}}$$

$$= \arg \max_\theta \frac{P(x > \theta|c_1)}{\beta[P(x > \theta|c_1) + \frac{P_{c_0}}{P_{c_1}}P(x > \theta|c_0)] + (1 - \beta)}.$$

GATS is a general method that can be tuned according to different performance requirements, by employing the different optimization criteria. By employing F_β, GATS will adjust the threshold automatically according to the certain performance requirement, by setting a proper value of β. A bigger β gives a heavier weight for precision and will lead to a precision-oriented system and vice versa. The effects of β variation on performance monitoring will be discussed in detail in Section 7.3.

7.2.4 Implementation issues

There are several implementation issues for GATS. The first issue is how GATS should be combined with novelty mining. Figure 7.3 shows the flowchart of novelty mining combined with GATS. After predicting the ith document d_i, $i = 1, 2, \ldots$, the system checks whether there is any new user feedback for the current document or any history document. If there is any available new feedback, the current threshold will be updated by GATS. Finally, the system will use this newly updated threshold to predict the next incoming document.

The second implementation issue concerns whether the number of feedbacks is enough for Gaussian probability estimation in GATS, in the initial stage of novelty mining. In our experiments, we found that the minimum number of feedbacks n_{min} of both novel and nonnovel documents should not be less than 4. A smaller n_{min} will degrade the accuracy of probability estimation and lead to an unreliable Gaussian probability model, while a large n_{min} will not start the adaptive threshold setting until the system accumulates enough user feedback. In our experimental study, we set $n_{min} = 4$ for both novel and nonnovel documents.

When the number of feedbacks does not meet the requirement of n_{min}, the initial threshold is necessary. Therefore, setting the initial threshold is another implementation issue. Due to the characteristics of novelty mining, we found that there are more novel documents in the early stage of document accumulation. Therefore, the initial threshold should be a little lower, where most of the documents can be retrieved. As the accumulating documents increase, the possible user feedback also increases to trigger GATS. In our experiments, we set the initial threshold $\theta_0 = 0.3$ for novelty scores.

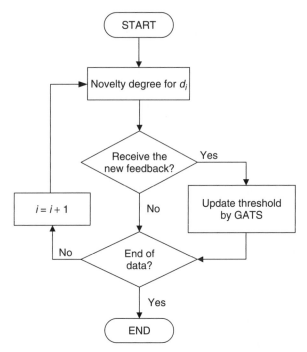

Figure 7.3   Novelty mining combined with GATS.

7.3   Experimental study

7.3.1   Datasets

Two public datasets, TREC 2004 Novelty Track data (Soboroff 2004) and TREC 2003 Novelty Track data (Soboroff and Harman 2003), were used in our experiments. The TREC 2004 and 2003 Novelty Track data is developed from AQUAINT collection. The news providers of the document set are Xin Hua, New York Times, and APW. This data is for sentence-level novelty mining, where both relevant and novel sentences for all 50 topics are selected by TREC's assessors and retrieved from the National Institute of Standards and Technology (NIST). For TREC 2004, there were a total of 8343 relevant sentences, of which 3454 (41.4%) were novel. In TREC 2003, 10 226 (65.73%) out of 15 557 sentences were novel.

From the TREC 2004 and 2003 Novelty Track sentence-level data, we built a set of document-level datasets, document-level TREC 2004 and document-level TREC 2003. In order to obtain the documents, we first combined the sentences by sentence type (headline or text), into documents according to their document id. Then, we performed our experiments on the document-level TREC 2004/2003. Because we already had the ground truth for the novelty of each TREC sentence, we easily calculated the actual percentage of novel sentences (PNS) in that document. If we set a low PNS threshold, most documents in the dataset are considered to be novel. By choosing to set different thresholds, we can observe the performance of GATS document-level novelty mining on datasets with different PNS.

In this experimental study, the focus was on novelty mining rather than relevant document categorization. Therefore, our experiments start with all given relevant documents (sentences), from which the novelty documents (sentences) are identified.

7.3.2 Working example

To illustrate the use of GATS in practice, we will first present a working example of GATS used for sentence-level novelty mining. Consider the following sentences from topic N39 in TREC 2003:

1. CLUES POINT TO PHILIPPINE STUDENT AS VIRUS AUTHOR By JOHN MARKOFF c.2000 N.Y. Times News Service

2. Law enforcement officials and computer security investigators focused on the Philippines Friday in their search for the author of a software program that convulsed the world's computer networks.

3. Investigators in both Asia and the United States said clues appeared to point to a college student in his early 20s using a Philippine Internet service provider.

4. The rogue program, borne as an attachment to an e-mail with the subject line 'I Love You,' surfaced in Asia on Wednesday.

5. It moved from there to Europe and the United States on Thursday, clogging or disabling corporate e-mail systems and destroying data on personal computers.

6. Although the spread of the infection appeared to slow Friday, at least eight variations of the original program had been identified by antivirus firms.

7. Once it is launched, the 'I Love You' program, among other things, tries to fetch an additional program from a Philippine Web site enabling it to steal passwords from the victim's computer.

8. American security experts said they had found evidence that a person using the 'spyder' alias found in the 'I Love You' program had written two versions of a password-stealing program found in recent months.

9. 'Our theory is that he had written this program twice and was looking for a way to get broader distribution for it,' said Peter S. Tippett of ICSA.net, a computer security firm based in Reston, Va.

10. At the same time, Fredrik Bjorck, a Swedish computer security researcher who last year helped identify the author of a similar program called

Melissa, told Swedish television that he had identified the perpetrator of the latest attack as a German exchange student named Mikael.

11. He said that Mikael was in his 20s and that he had used Internet service providers in the Philippines to spread his programs.

12. Bjorck said Mikael had published information on how to get rid of the 'I Love You' program.

13. He did not identify Mikael's location.

14. The ICSA.net researchers said they had disassembled one of the four components of the 'I Love You' program and had discovered that its instructions closely matched two similar programs that they had captured last fall and in January.

15. Once a computer was infected, the program was set up to fetch the password-stealing component from a Philippine Web site.

16. After it was installed in the computer it was programmed to relay the stolen passwords to an e-mail account also in the Philippines.

17. But after the 'I Love You' outbreak was detected on Wednesday, the company running the Philippine Web site, Sky Internet, quickly removed the password program from its system.

18. Computer investigators said that both the 'I Love You' program and the password-stealing modules discovered earlier had references to Amable Mendoza Aguiluz Computer College, which they said had seven campuses in the Philippines.

When sentence-level novelty mining is used with a fixed threshold of 0.55, sentences 9, 11, and 15 are determined as nonnovel, as shown in Figure 7.4 (a cross is nonnovel, a check novel). If we provide feedback as shown and process the sentences again using the GATS option, then the threshold will be automatically adjusted. We set the feedback at '1' for 'Novel' for sentences 2–5, and '0' for 'Nonnovel' for sentences 8, 9, 11, and 15, as shown in Figure 7.4. In this scenario, after running GATS based on user feedback, the threshold was automatically adjusted to 0.60 for sentences 16 and beyond, as shown in Figure 7.5. In this figure, sentence 16 was compared to the most similar sentence, in this case sentence 15, and because the novelty score of 0.5980 is below the threshold value, sentence 16 is now rated as 'Novel'. As seen in Figure 7.6, the resulting novelty rating changed for sentences 16, 17, and 18 from 'Novel' to 'Nonnovel', based on the threshold adjustment from user feedback. This example shows how GATS works for a real-life scenario.

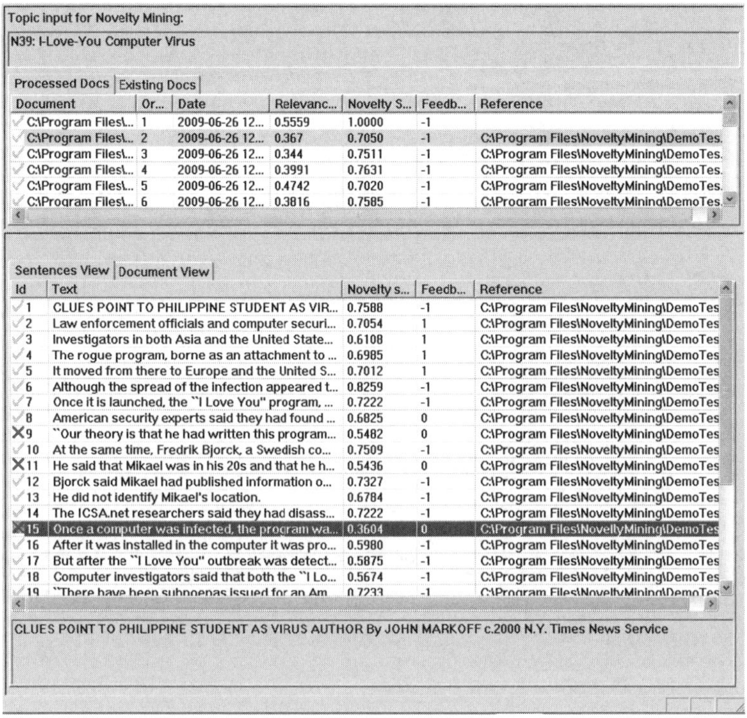

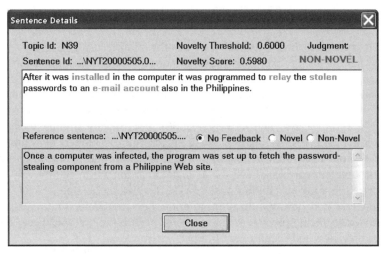

Figure 7.4   Sentence-level novelty mining results for TREC03 topic N39.

Figure 7.5   Threshold adjustment to 0.6000 for sentence 16 after running GATS.

Topic input for Novelty Mining:

N39: I-Love-You Computer Virus

Processed Docs | Existing Docs

Document	Or...	Date	Relevanc...	Novel!
✓ C:\Program Files\NoveltyMining\DemoTest\N39\NYT20000505.0030.txt	1	2009-06-26 12...	0.5559	1.0000
C:\Program Files\NoveltyMining\DemoTest\N39\NYT20000505.0340.txt	2	2009-06-26 12...	0.367	0.7050
✓ C:\Program Files\NoveltyMining\DemoTest\N39\NYT20000510.0216.txt	3	2009-06-26 12...	0.344	0.7511
✓ C:\Program Files\NoveltyMining\DemoTest\N39\NYT20000510.0230.txt	4	2009-06-26 12...	0.3991	0.7631
✓ C:\Program Files\NoveltyMining\DemoTest\N39\NYT20000511.0127.txt	5	2009-06-26 12...	0.4742	0.7020
✓ C:\Program Files\NoveltyMining\DemoTest\N39\NYT20000517.0178.txt	6	2009-06-26 12...	0.3816	0.7585

Sentences View | Document View

Id	Text	Novelty s...	Feedb...	Reference
✓ 1	CLUES POINT TO PHILIPPINE STUDENT AS VIR...	0.7588	-1	C:\Program Files\NoveltyMining\DemoTes
✓ 2	Law enforcement officials and computer securi...	0.7054	1	C:\Program Files\NoveltyMining\DemoTes
✓ 3	Investigators in both Asia and the United State...	0.6108	1	C:\Program Files\NoveltyMining\DemoTes
✓ 4	The rogue program, borne as an attachment to ...	0.6985	1	C:\Program Files\NoveltyMining\DemoTes
✓ 5	It moved from there to Europe and the United S...	0.7012	1	C:\Program Files\NoveltyMining\DemoTes
✓ 6	Although the spread of the infection appeared t...	0.8259	-1	C:\Program Files\NoveltyMining\DemoTes
✓ 7	Once it is launched, the ``I Love You'' program, ...	0.7222	-1	C:\Program Files\NoveltyMining\DemoTes
✗ 8	American security experts said they had found ...	0.6825	0	C:\Program Files\NoveltyMining\DemoTes
✗ 9	``Our theory is that he had written this program...	0.5482	0	C:\Program Files\NoveltyMining\DemoTes
✓ 10	At the same time, Fredrik Bjorck, a Swedish co...	0.7509	-1	C:\Program Files\NoveltyMining\DemoTes
✗ 11	He said that Mikael was in his 20s and that he h...	0.5436	0	C:\Program Files\NoveltyMining\DemoTes
✓ 12	Bjorck said Mikael had published information o...	0.7327	-1	C:\Program Files\NoveltyMining\DemoTes
✓ 13	He did not identify Mikael's location.	0.6784	-1	C:\Program Files\NoveltyMining\DemoTes
✓ 14	The ICSA.net researchers said they had disass...	0.7222	-1	C:\Program Files\NoveltyMining\DemoTes
✗ 15	Once a computer was infected, the program wa...	0.3604	0	C:\Program Files\NoveltyMining\DemoTes
✗ 16	After it was installed in the computer it was pro...	0.5980	-1	C:\Program Files\NoveltyMining\DemoTes
✗ 17	But after the ``I Love You'' outbreak was detect...	0.5875	-1	C:\Program Files\NoveltyMining\DemoTes
✗ 18	Computer investigators said that both the ``I Lo...	0.5674	-1	C:\Program Files\NoveltyMining\DemoTes
✓ 19	``There have been subpoenas issued for an Am...	0.7233	-1	C:\Program Files\NoveltyMining\DemoTes

After it was installed in the computer it was programmed to relay the stolen passwords to an e-mail account also in the Philippines.

Figure 7.6   Sentence-level novelty mining results for TREC03 topic N39 after running GATS.

7.3.3   Experiments and results

We also compared the performance of novelty mining using GATS to that of novelty mining using fixed thresholds. Figure 7.7 shows the precision–recall (PR) curves of these two algorithms on TREC 2004 Novelty Track data. In information retrieval, PR curves are commonly used to compare algorithms, where the algorithm with a larger area under the curve is regarded as a better algorithm (Davis and Coadrich 2006). For novelty mining using fixed thresholds, the corresponding PR curve (black line) is plotted by varying the fixed threshold from 0.05 to 0.95. For each threshold, the precision and recall for each topic are calculated and the average precision and recall over 50 topics are reported. For novelty mining with GATS, the PR curve is plotted by varying the parameter β of the optimization criterion F_β score from 0.1 to 0.9. Again, for each value β, the precision and recall for each topic are calculated and the average precision and recall over 50 topics are reported.

From Figure 7.7, we can observe that novelty mining with GATS outperforms the system with fixed thresholds. The precision and recall obtained by novelty

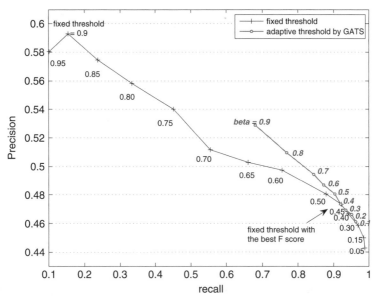

Figure 7.7   Precision–recall curves of novelty mining with fixed threshold vs. adaptive threshold by GATS (tuning for F_β) with complete user feedback on TREC 2004 Novelty Track data.

mining with GATS will not fall within the regions of the extreme values, in which the F score can be very low. In practice, our users usually require a high-recall system with the precision no lower than a lower bound, or a high-precision system with the recall no lower than a lower bound. An extremely high recall with an extremely low precision is useless because this system just marks almost all documents as novel. On the other hand, an extremely high precision with an extremely low recall means that the system only marks very few documents as novel. Both cases make little sense.

Moreover, since there is no prior information available for a user to choose a suitable fixed threshold, the system with a predefined threshold can hardly lead to a suitable tradeoff between precision and recall, and hence can hardly obtain a good F score. On the contrary, GATS will optimize the F score automatically, based on user feedback over time.

Besides the PR curve, we also compare two algorithms using the F_β score. Table 7.2 shows the performance of the two algorithms evaluated with F_β scores of $\beta = 0.2$, 0.5, and 0.8. For novelty mining employing GATS, the parameter β is set to 0.2, 0.5, and 0.8 accordingly. For novelty mining employing the fixed threshold, the highest F_β scores are reported in tables after various trial-and-error attempts. From Table 7.2, by comparing to the best fixed threshold, we discovered that GATS can obtain similar or slightly better results for TREC

Table 7.2   Comparison of performance evaluated by F_β
($\beta = 0.2, 0.5, 0.8$) on TREC 2004 Novelty Track data.

	Performance of the novelty mining system	
	Adaptive threshold by GATS (β)	Best fixed threshold by trial and error (θ)
$F_{0.2}$	0.7706 (0.2)	0.7758 (0.15)
$F_{0.5}$	0.6155 (0.5)	0.6126 (0.45)
$F_{0.8}$	0.5396 (0.8)	0.5281 (0.60)

2004 Novelty Track data. The best fixed thresholds for $F_{0.2}$, $F_{0.5}$, and $F_{0.8}$ are 0.15, 0.45, and 0.60, respectively. Examination of the PR curves in Figure 7.7 suggests that the corresponding region of the best fixed thresholds is covered by the PR curve of GATS. This implies that GATS can be effective in searching for the best threshold in novelty mining, under different performance requirements.

In the following subsections, we test GATS by assuming complete feedback for document-level novelty mining (NM) data with low, medium, and high novelty ratios. This will provide some guidelines on how GATS should be used.

Case 1: High novelty ratio

To construct document-level NM data with a high novelty ratio, we chose TREC 2003 Novelty Track data because the ground truth novelty ratio at sentence level was naturally high (65.73%). By setting the PNS threshold to 0.25, the document-level novelty ratio is 79.20%, i.e. 79.20% of incoming documents are novel. In this case, GATS does not perform as well as the best result of the fixed threshold (see Figure 7.8).

Case 2: Medium novelty ratio (30%–75%)

We constructed document-level NM data with a medium novelty ratio by using TREC 2004 Novelty Track data because the ground truth novelty ratio at sentence level is 41.40%. By setting the PNS threshold to 0.03, the document-level novelty ratio is 47.73%, i.e. 47.73% of incoming documents are novel. In this case, GATS performs comparable to the best result of the fixed threshold (see Figure 7.9).

Case 3: Low novelty ratio (30%)

We also constructed document-level NM data with a low novelty ratio by using TREC 2004 Novelty Track data. In setting the PNS threshold to 0.5, the document-level novelty ratio is 27.71%, i.e. 27.71% of incoming documents are novel. In this case, the GATS algorithm outperforms the best result of the fixed threshold (see Figure 7.10).

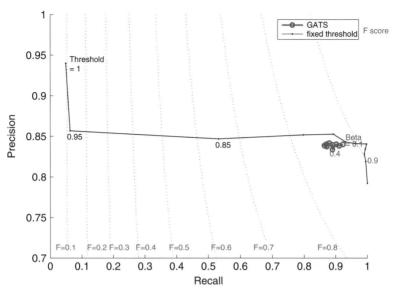

Figure 7.8   Precision–recall curves of novelty mining with fixed threshold vs. adaptive threshold by GATS (tuning for F_β) with complete user feedback on document-level TREC 2003 Novelty Track data (with PNS threshold 0.25).

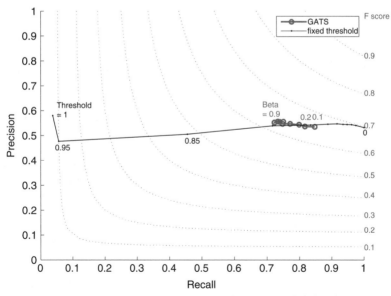

Figure 7.9   Precision–recall curves of novelty mining with fixed threshold vs. adaptive threshold by GATS (tuning for F_β) with complete user feedback on document-level TREC 2004 Novelty Track data (with PNS threshold 0.03).

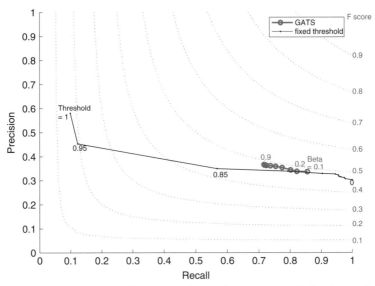

Figure 7.10   Precision–recall curves of novelty mining with fixed threshold vs. adaptive threshold by GATS (tuning for F_β*) with complete user feedback on document-level TREC 2004 Novelty Track data (with PNS threshold 0.5).*

Discussion

Although both the fixed threshold and the GATS parameter β control the trade-off between precision and recall, they play different roles in novelty mining. The fixed threshold cannot reflect the trade-off between precision and recall directly. Since different data may have different characteristics and different metrics may output different values of novelty scores, the fixed threshold can hardly be predefined with confidence. On the contrary, the parameter β in GATS reflects the weights of precision and recall directly (β is the weight of precision while $1 - \beta$ is the weight of recall), and hence can be set based on the performance requirement directly.

From our experimental results on document-level NM data with low, medium, and high novelty ratios, we find that GATS is extremely useful for data with low novelty ratios, useful for data with medium novelty ratios, but not as useful as the best fixed threshold for data with a novelty ratio higher than 75%. Therefore, GATS is not recommended for topics with high novelty ratios. In this case, setting a lower fixed threshold to force most of the documents to be 'novel' would be a better choice.

7.4   Conclusion

This chapter addressed the problem of setting an adaptive threshold by utilizing user feedback over time. The proposed method, the Gaussian-based adaptive threshold setting (GATS) algorithm, modeled the distributions of novelty scores from both novel and nonnovel classes by the Gaussian distributions. Class distributions learnt from user feedback yielded the global information of data used for the construction of an optimization criterion for searching the best threshold. GATS is a general method, which can be tuned according to different performance requirements, by combining with different optimization criteria. In this chapter, the most commonly used performance evaluation measure in NM, the F_β score, has been employed as the optimization criterion. The F_β score is the weighted harmonic average of precision and recall, where β and $(1 - \beta)$ are weights for precision and recall, respectively.

In the experimental study, the NM system employing the GATS algorithm was tested on experimental datasets with complete user feedback on data with low, medium, and high novelty ratios (percentage of novel sentences/documents). The experimental results suggest that GATS is very effective in finding the best threshold in the NM system. Moreover, GATS is able to meet the different performance requirements by setting the weights of precision and recall externally. GATS has been shown to be extremely effective for data with a low novelty ratio, useful for data with a medium novelty ratio, and not as effective for data with a high novelty ratio.

References

Allan J, Wade C and Bolivar A 2003 Retrieval and novelty detection at the sentence level. *SIGIR 2003, Toronto, Canada*, pp. 314–321 ACM.

Arampatzis A, Beney J, Koster CHA and Weide TP 2000 KUN on the TREC-9 filtering track: Incrementality, decay, and threshold optimization for adaptive filtering systems. *TREC 9 – the 9th Text Retrieval Conference*.

Davis J and Coadrich M 2006 The relationship between precision-recall and ROC curves. *Proceedings of the 23rd International Conference on Machine Learning*, pp. 233–240.

Harman D 2002 Overview of the TREC 2002 Novelty Track. *TREC 2002 – the 11th Text Retrieval Conference*, pp. 46–55.

Kwee AT, Tsai FS and Tang W 2009 Sentence-level novelty detection in English and Malay. *Lecture Notes in Computer Science (LNCS)* vol. 5476 Springer pp. 40–51.

Ng KW, Tsai FS, Goh KC and Chen L 2007 Novelty detection for text documents using named entity recognition. *6th International Conference on Information, Communications and Signal Processing*, pp. 1–5.

Robertson S 2002 Threshold setting and performance optimization in adaptive filtering. *Information Retrieval* **5**(2–3), 239–256.

Soboroff I 2004 Overview of the TREC 2004 Novelty Track. *TREC 2004 – the 13th Text Retrieval Conference*.

Soboroff I and Harman D 2003 Overview of the TREC 2003 Novelty Track. *TREC 2003 – the 12th Text Retrieval Conference*.

Tang W and Tsai FS 2009 Intelligent novelty mining for the business enterprise. *Technical Report*.

Zhai C, Jansen P, Stoica E, Grot N and Evans DA 1999 Threshold calibration in CLARIT adaptive filtering. *Proceedings of the Seventh Text Retrieval Conference, TREC-7*, pp. 149–156.

Zhang Y and Callan J 2001 Maximum likelihood estimation for filtering thresholds. *ACM SIGIR 2001*, pp. 294–302.

Zhang Y and Tsai FS 2009a Chinese novelty mining *EMNLP'09: Proceedings of the Conference on Empirical Methods in Natural Language Processing*, pp. 1561–1570.

Zhang Y and Tsai FS 2009b Combining named entities and tags for novel sentence detection. *ESAIR'09: Proceedings of the WSDM'09 Workshop on Exploiting Semantic Annotations in Information Retrieval*, pp. 30–34.

Zhang Y, Callan J and Minka T 2002 Novelty and redundancy detection in adaptive filtering. *ACM SIGIR 2002, Tampere, Finland*, pp. 81–88.

Zhao L, Zheng M and Ma S 2006 The nature of novelty detection. *Information Retrieval* **9**, 527–541.

8

Text mining and cybercrime

April Kontostathis, Lynne Edwards and Amanda Leatherman

8.1 Introduction

According to the most recent 2008 online victimization research, approximately 1 in 7 youths (ages 10 to 17 years) experience a sexual approach or solicitation by means of the Internet (National Center for Missing and Exploited Children 2008). In response to this growing concern, law enforcement collaborations and nonprofit organizations have been formed to deal with sexual exploitation on the Internet. Most notable is the Internet Crimes Against Children (ICAC) task force (Internet Crimes Against Children 2009). The ICAC Task Force Program was created to help state and local law enforcement agencies enhance their investigative response to offenders who use the Internet, social networking websites, or other computer technology to sexually exploit children. The program is currently composed of 59 regional task force agencies and is funded by the United States Department of Justice, Office of Juvenile Justice and Delinquency Prevention.

The National Center for Missing and Exploited Children (NCMEC) has set up a CyberTipLine for reporting cases of child sexual exploitation including child pornography, online enticement of children for sex acts, molestation of children outside the family, sex tourism of children, child victims of prostitution, and unsolicited obscene material sent to a child. All calls to the tip line are referred to appropriate law enforcement agencies – and the magnitude of the calls is staggering. From March 1998, when the CyberTipLine began operations, until April 20, 2009, there were 44 126 reports of 'Online Enticement of Children for

Text Mining: Applications and Theory   edited by Michael W. Berry and Jacob Kogan
© 2010, John Wiley & Sons, Ltd

Sexual Acts', one of the reporting categories. There were 146 in the week of April 20th, 2009 alone (National Center for Missing and Exploited Children 2008).

The owners of Perverted-Justice.com (PJ) began a grassroots effort to identify cyberpredators in 2002. PJ volunteers pose as youths in chat rooms and respond when approached by an adult seeking to begin a sexual relationship with a child. We are currently working with the data collected by PJ from these conversations in an effort to understand cyberpredator communications.

Cyberbullying, according to the National Crime Prevention Council, is using the Internet, cell phones, video game systems, or other technology to send or post text or images intended to hurt or embarrass another person – and is a growing threat among children. In 2004, half of US youths surveyed stated that they or someone they knew had been victims or perpetrators of cyberbullying (National Crime Prevention Council 2009a). Being a victim of cyberbullying is a common and painful experience. Nearly 20% of teens had a cyberbully pretend to be someone else in order to trick them online, getting the victim to reveal personal information; 17% of teens were victimized by someone lying about them to others online; 13% of teens learned that a cyberbully was pretending to be them while communicating with someone else; and 10% of teens were victimized by someone posting unflattering pictures of them online, without permission (National Crime Prevention Council 2009b).

The anonymous nature of the Internet may contribute to the prevalence of cyberbullying. Kids respond to cyberbullying by avoiding communication technologies or messages altogether. They rarely report the conduct to parents (for fear of losing phone/Internet privileges) or to school officials (for fear of getting into trouble for using cell phones or the Internet in class) (Agatston et al. 2007; Williams and Guerra 2007).

As we analyzed cyberbullying and cyberpredator transcripts from a variety of sources, we were struck by the similar communicative tactics employed by both cyberbullies and cyberpredators – in particular, masking identity and deception. We were also struck by the similar responses of law enforcement and youth advocacy groups: reporting and preventing. Victims are physically and psychologically abused by predators and bullies who trap them in vicious communicative cycles using modern technologies; their only recourse is to report the act to authorities after it has occurred. By the time a report is made, unfortunately the aggressor has moved on to a new victim.

Cyberbullying and Internet predation frequently occur over an extended period of time and across several technological platforms (i.e. chat rooms, social networking sites, cell phones, etc.). Techniques that link multiple online identities would help law enforcement and national security agencies identify criminals, as well as the forums in which they participate. The threat to youth is of particular interest to researchers, law enforcement, and youth advocates because of the potential for it to get worse as membership of online communities continues to grow (Backstrom et al. 2006; Kumar et al. 2004; Leskovec et al. 2008) and as new social networking technologies emerge (Boyd and Ellison 2007). Much of modern communication takes place via online chat media in virtual communities populated by millions of anonymous members who use a variety of chat technologies to maintain virtual relationships based on daily (if not hourly) contact (Ellison et al. 2007; O'Murchu et al. 2004). MSN Messenger, for example, reports 27 million users and AOL Instant Messenger has the largest share of the instant messaging market (52% as of 2006) (IM MarketShare 2009); however, Facebook, the latest social networking craze, reported over 90 million users worldwide (Nash 2008). These media, along with MySpace, WindowsLive, Google, and Yahoo, all have online chat technologies that can be easily accessed by anyone who chooses to create a screen name and to log on; no proof of age, identity, or intention is required. A recent update to Facebook also allows users to post and receive Facebook messages via text messaging on their cell phones (FacebookMobile 2009).

We describe the current state of research in the areas of cyberbullying and Internet predation in Section 8.2. In Section 8.3, we describe several commercial products which claim to provide chat and social networking site monitoring for home use. Finally in Section 8.4 we offer our conclusions and discuss opportunities for future research into this interesting and timely field.

8.2 Current research in Internet predation and cyberbullying

This section provides a summary of research into Internet predation and cyberbullying. We first review the technology that is available for capturing Internet Messenger (IM) and Internet Relay Chat (IRC). Next we discuss the datasets that are currently available for research in the area. Finally we survey several research articles for both Internet predation and cyberbullying detection, as well as provide a summary of the literature as it relates to legal issues.

8.2.1 Capturing IM and IRC chat

Data collection is the first step in any research project in text mining. Data collection for the study of cybercrime needs to focus primarily on capturing data from chat rooms and social networking sites; however, there are both legal and technical issues that must be overcome. In this section we discuss the work by several research groups which have successfully captured online chat.

In Dewes et al. (2003) a multi-layered approach for capturing web chat from various sources including IRC and Web-based (both HTTP and java) chat systems is used. They begin by casting a wide net, essentially capturing all network traffic that passes through a particular router. Several filters are then applied to separate the chat traffic from nonchat traffic. Early experiments show that 91.7% of the chat traffic can be identified (recall) and 93.7% of the traffic that is captured is indeed chat (precision).

Other research groups take a more direct approach. Gianvecchio et al. signed into Yahoo chat rooms and logged all posts for a two-week period in order

to capture data for their bot detection study (Gianvecchio et al. 2008). Others
set up host servers and monitor all activity directly at the server level (Cooke
et al. 2005). Several low-cost commercial products for capturing relevant network
packets are also available (ICQ-Sniffer 2009).

8.2.2 Current collections for use in analysis

There is very little reliable labeled data concerning predator communications;
much of the work that has appeared in both computer science and communication studies forums is focused on anecdotal evidence and chat log transcripts
from PJ (Perverted-Justice.com 2008). PJ began as a grassroots effort to identify
cyberpredators. Its volunteers pose as youths in chat rooms and respond when
approached by an adult seeking to begin a sexual relationship with a minor.
When these activities result in an arrest and conviction, the chat log transcripts
are posted online. New chat logs continue to be added to the website. There
were 325 transcripts, representing arrests and convictions, on the site as of July
2009. Details about early research projects that use this data are described in
Section 8.2.4.

The use of PJ transcripts for research into cyberpredation is controversial. The
logs contain transcripts of conversations between a predator and a pseudo-victim,
an adult posing as a young teenager. However, the predators who participated
in these conversations were convicted, based, at least in part, on the content
of the chat logs, which provides a measure of credibility to the data. We have
been in communication with several researchers who are working on related
projects in computer science, media and communication studies, criminal justice,
and sociology and have not been able to identify another source of data. We
will continue to seek transcripts that contain conversations between predators
and minors; however, it will be extremely difficult. Law enforcement agencies
are rarely able to share chat log transcripts (when they have them), even for
scholarly examination, because the logs are not stored in a central repository and
only excerpts are used when cases go to trial (Personal Communication 2008).

A second dataset was created by Dr Susan Gauch, University of Arkansas,
who collected chat logs during a chat room topic detection project (Bengel
et al. 2004). Dr Gauch's project included the development of a crawler that
downloaded chat logs (ChatTrack). Unfortunately, the software is no longer
available. This chat data, although somewhat dated, has been used in some
of the preliminary studies involving an analysis of predator communications
(Kontostathis et al. 2009).

We have identified one additional publically available dataset which can
be used for research on the communication styles of cybercriminals. In 2009,
the Content Analysis for the Web 2.0 workshop (held in conjunction with
WWW2009) proposed three independent shared tasks: text normalization,
opinion and sentiment analysis, and misbehavior detection. The misbehavior
detection task addressed the problems of detecting inappropriate activity in
which some users in a virtual community are harassing or offensive to some

other members of the community. A common training dataset was made available to all task participants. The provided dataset was intended as a representative sample of what can be found in Web 2.0. The data were collected from five different public sites, including Twitter, MySpace, Slashdot, Ciao, and Kongregate. Interested parties should refer to the CAW 2.0 website for additional information (CAW2.0 2009). This data is exclusively intended for research purposes. A research project which used this data to detect cyberbullying is discussed in Section 8.2.5.

8.2.3   Analysis of IM and IRC chat

Much of the social networking research in computer science has focused on chat room data (Jones et al. 2008; Muller et al. 2003). A lot of this work has centered on identifying discussion thread subgroups within a chat forum (Acar et al. 2005; Camtepe et al. 2004); some researchers focus on the technical difficulties encountered when trying to parse chat log data (Tuulos and Tirri 2004; Van Dyke et al. 1999). Surprisingly few researchers have attempted to deal with the creation of specific applications for analysis and management of Internet predators or cyberbullies. The few that we have identified are described in the following subsections.

8.2.4   Internet predation detection

We have identified articles that take two different approaches to detection of cyberpredator communications. The first uses a bag-of-words approach and a standard statistical classification technique. The second leverages research in communications theory to develop more sophisticated features for input to the classifier.

A statistical approach

Pendar used the PJ transcripts to separate predator communication from victim communication (Pendar 2007). In this study, the author downloaded the PJ transcripts and indexed them. After preprocessing to reduce some of the problems associated with Internet communication (i.e. handling netspeak), the author developed attributes for each chat log. The attributes consisted of word unigrams, bigrams, and trigrams. Terms that appeared in only one log or in more than 95% of the logs were removed from the index. Afterward approximately 10 000 unigrams, 43 000 bigrams, and 13 000 trigrams remained. The author describes using 701 log files.[1]. Each log file was split into victim communication and predator communication, resulting in 1402 total input instances, each with 10 000–43 000 attributes, depending on the model being tested. Additional feature extraction and weighting completed the indexing process.

---

[1] It appears as if the perverted-justice.com site has changed its method of presenting the chat data in recent years.

The data file was split into a 1122 instance training set and a 280 instance test set, stratified by class (i.e. the test set contained 140 predator instances and 140 victim instances). Classification was then attempted using both support vector machine (SVM) and distance-weighted k-nearest neighbor (k-NN) classifiers. The F-measure (see also Sections 3.4 and 7.2.3) reported by the author ranged from 0.415 to 0.943. The k-NN classifier was a better classifier for this task and trigrams were shown to be more effective than unigrams and bigrams. The maximum performance (F-measure $= 0.943$) was obtained when 30 nearest neighbors were used and 10 000 trigrams were extracted and used as attributes.

An approach based on communicative theory

In contrast to the purely statistical methods employed by Pendar, Kontostathis et al. used a rule-based approach in Kontostathis et al. (2009). This project integrates communication and computer science theories and methodologies to develop tools to protect children from cyberpredators.

The theory of luring communication provides a model of the communication processes that child sexual predators use in the real world to entrap their victims Olson et al. (2007). This model consists of three major stages:

1. gaining access to the victim;

2. entrapping the victim in a deceptive relationship;

3. initiating and maintaining a sexually abusive relationship.

During the gaining access phase, the predator maneuvers him- or herself into professional and social positions where he or she can interact with the child in a seemingly natural way, while still maintaining a position of authority over the child. For example, gaining employment at an amusement park or volunteering for a community youth sports team. The next phase, entrapping the victim in a deceptive relationship, is a communicative cycle that consists of grooming, isolation, and approach. Grooming involves subtle communication strategies that desensitize victims to sexual terminology and reframe sexual acts in child-like terms of play or practice. In this stage, offenders also isolate their victims from family and friend support networks before approaching the victim for the third phase: sexual contact and long-term abuse.

In previous work, we expanded and modified the luring theory to accommodate the difference between online luring and real-world luring (Leatherman 2009). For example, the concept 'gaining access' was revised to include the initial entrance into the online environment and initial greeting exchange by offenders and victims, which is different from meeting kids at the amusement park or through a youth sports league. Communicative desensitization was modified to include the use of slang, abbreviations, netspeak, and emoticons in online conversations. The core concept underpinning entrapment is the ongoing deceptive trust that develops between victims and offenders. In online luring communications, this concept is defined as perpetrator and victim sharing personal information, information about activities, relationship details, and compliments.

Communications researchers define two primary goals for content analysis (Riffe et al. 1998):

1. describe the communication; and

2. draw inferences about its meaning.

In order to perform a content analysis for Internet predation, we developed a codebook and dictionary to distinguish among the various constructs defined in the luring communication theoretical model. The coding process occurred in several stages. First, a dictionary of luring terms, words, icons, phrases, and net-speak for each of the three luring communication stages was developed. Second, a coding manual was created. This manual has explicit rules and instructions for assigning terms and phrases to their appropriate categories. Finally, software that mimics the manual coding process was developed (this software is referred to as *ChatCoder* below).

Twenty-five transcripts from the PJ website were carefully analyzed for the development of the dictionary. These 25 online conversations ranged from 349 to 1500 lines of text. The perpetrators span from 23 to 58 years of age, were all male, and were all convicted of sexual solicitation of minors over the Internet.

We captured key terms and phrases that were frequently used by online sexual predators, and identified their appropriate category labels within the luring model: deceptive trust development, grooming, isolation, and approach (Leatherman 2009; Olson et al. 2007). The dictionary included terms and phrases common to net culture in general, and luring language in particular. Some examples appear in Table 8.1. The version of coding dictionary used in these experiments contained 475 unique phrases. A breakdown of the phrase count by category appears in Table 8.2.

In order to provide a baseline for the usefulness of the codebook for detecting online predation, we ran two small categorization experiments. In the first experiment, we coded 16 transcripts in two ways: first we coded the predator dialogue (so only phrases used by the predator were recorded), and then we coded for the victim. Thus, we had 32 instances, and each instance had a count of the phrases in each of the coding categories (eight attributes). Our class attribute was binary (predator or victim).

We used the J48 classifier within the Weka suite of data mining tools (Witten and Frank 2005) to build a decision tree to predict whether the coded dialogue was predator or victim. The J48 classifier builds a C4.5 decision tree with reduced error pruning (Quinlan 1993). This experiment is similar to that in Pendar (2007), but Pendar used a bag-of-words approach and an instance-based learner. The classifier correctly predicted the class 60% of the time, a slight improvement over the 50% baseline. This is remarkable when we consider the fact that we were coding individuals who were in conversation with each other, and therefore the

Table 8.1   Sample excerpts from the codebook for Internet predation.

Phrase	Coding category
are you safe to meet	Approach
i just want to meet	Approach
i just want to meet and mess around	Approach
how cum	Communicative desensitization
if i don't cum right back	Communicative desensitization
i want to cum down there	Communicative desensitization
i just want to gobble you up	Communicative desensitization
you are a really cute girl	Compliment
you are a sweet girl	Compliment
are you alone	Isolation
do you have many friends	Isolation
let's have fun together	Reframing
let's play a make believe game	Reframing
there is nothing wrong with doing that	Reframing

Table 8.2   Dictionary summary - phrase count by category.

Category	Phrase count
Activities	11
Approach	56
Communicative desensitization	220
Compliment	35
Isolation	43
Personal information	29
Reframing	57
Relationship	24

terminology used was similar. Stratified threefold cross-validation, as implemented within Weka, was used to evaluate the results.

In a second experiment we built a C4.5 decision tree to distinguish between PJ and ChatTrack transcripts. The ChatTrack dataset is described in Section 8.2.2. We coded 15 PJ transcripts (both victim and predator dialogue) and 14 transcripts from the ChatTrack dataset (Bengel et al. 2004). The classifier that was built was able to distinguish the PJ transcripts 93% of the time. We also used stratified threefold cross-validation for evaluation in these experiments.

As we analyzed the PJ transcripts, we noticed recurring patterns within the dialogue used by the suspects and began to wonder if we could cluster different types of predators via their language pattern usage.

We chose the k-means (Hartigan and Wong 1979) clustering algorithm because it is known to be both simple and effective. The k-means algorithm partitions a set of objects into k subclasses. It attempts to find the centers of natural clusters in the data by assuming that the object attributes form a vector space, and minimizing the intra-cluster variance. Thus, k-means generally forms tight, circular clusters around a centroid, and the algorithm outputs this centroid. k-means is particularly applicable to numeric attributes, and all of our attributes are numeric.

In our experiments, we counted the number of phrases in each of the eight coding categories for the 288 transcripts that were available on the PJ website as of August 2008 (predator only), and created an eight-dimensional vector for each instance. Thus, we used the same attributes that were used in the categorization experiments, but we were able to use all of the PJ transcripts. The vectors were column normalized by dividing by the maximum value in each column (i.e. all *activities* values were divided by the maximum value for *activities*). These vectors were then input to the k-means algorithm, and a set of clusters was determined.

The user must provide a value of k to the k-means clustering tool, and we were unsure about the number of categories of suspects that we might find, so we tried various values for k. We found that $k = 4$ produced the best result (the minimum intra-cluster variance), suggesting the hypothesis that there are four different types of Internet predators. More work is needed to determine labels for these categories of suspects. The centroid for each cluster appears in Figure 8.1. This figure clearly shows that some suspects spend more time overall with the victim (lines that are higher on the graph) and also that suspects in different clusters used different strategies during their conversations (as determined by line shape). For example, cluster 2 has a higher ratio of compliments vs. communicative desensitization as compared to cluster 3.

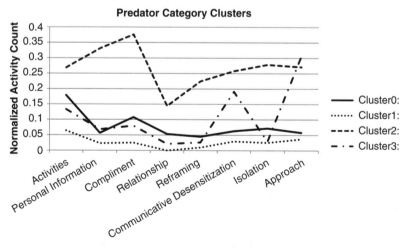

Figure 8.1   Initial clustering of predator type.

8.2.5   Cyberbullying detection

In 2006, the Conference on Human Factors in Computing Systems (CHI) ran a workshop on the misuse and abuse of interactive technologies, and in 2008 Rawn and Brodbeck showed that participants in first-person shooter games had a high level of verbal aggression, although in general there was no correlation between gaming and aggression (Rawn and Brodbeck 2008).

Most recently, in 2009 the Content Analysis for the Web 2.0 (CAW 2.0) workshop was formed and held in conjunction with WWW2009. As noted above, the CAW 2.0 organizers devised a shared task to deal with online harassment, and also developed a dataset to be used for research in this area. Only one submission was received for the misbehavior detection task. A brief summary of that paper follows.

Yin et al. define harassment as communication in which a user intentionally annoys another user in a web community. In Yin et al. (2009) detection of harassment is presented as a classification problem with two classes: positive class for posts which contain harassment and negative class for posts which do not contain harassment.

The authors combine a variety of methods to develop the attributes for input to their classifier. They use standard term weighting techniques, such as TFIDF (Term Frequency–Inverse Document Frequency) to extract index terms and give appropriate weight to each term. They also develop a rule-based system for capturing sentiment features. For example, a post that contains foul language and the word 'you' (which can appear in many forms in online communication) is likely to be an insult directed at someone, and therefore could be perceived as a bullying post. Finally, some web communities seem to engage in friendly banter or 'trash talk' that may appear to be bullying, but is instead just a communicative style. The authors also were able to identify contextual features by comparing a post to a window of neighboring posts. Posts that are unusual or which generate a cluster of similar activity from other users are more likely to be harassing.

After extracting relevant features, the authors developed an SVM classifier for detecting bullying behavior in three of the datasets provided by the CAW 2.0 conference organizers. They chose two different types of communities: Kongregate, which captures IM conversations during game play; and Slashdot/MySpace, which tend to be more asynchronous discussion-style forums where users write longer messages and discussion may continue over days or weeks. The authors manually labeled the three datasets. The level of harassment in general was very sparse. Overall only 42 of the 4802 posts in the Kongregate dataset represented bullying behavior. The ratio of bullying to nonbullying in Slashdot was similar (60 out of 4303 posts). MySpace was a little higher with 65 out of 1946 posts.

The authors employed an SVM to develop a model for classifying harassing posts. Their experimental results show that including the contextual and sentiment features improves the classification over the local weighting (TFIDF) baseline for the three datasets. The maximum recall was achieved with the chat-style collection (recall was 0.595 for Kongregate). Precision was best when the dataset contained more harassment (precision was 0.417 for MySpace). Overall the *F*-measure ranged from 0.298 to 0.442, so there is much room for improvement. A random chance baseline would be less than 1%, however, so the experimental results show that detection of cyberbullying is possible.

8.2.6 Legal issues

Companies have long been aware of the potential for misuse of email for bullying and harassment. In Sipior and Ward (1999), the authors report on the increased litigation surrounding sexual harassment in the workplace, particularly harassment via email.

Internet predation and cyberbullying are relatively new crimes, and, as such, the legal community is struggling to work with the technical community to protect victims while also protecting the civil rights of innocent users of Internet channels. Early attempts at collaboration between technicians and law enforcement, as described in a case study in Axlerod and Jay (1999), were initially frustrating. The collaborative work eventually paid off as computer scientists learned what is (and is not) permitted under the US legal system, and law enforcement officials learned to trust and use technical solutions to their best advantage.

In Burmester et al. (2005) the authors describe a combined hardware and software solution for providing law enforcement personnel with information in cases of cyberstalking. The article provides a profile of a technically advanced cyberstalker (who shares many traits with Internet predators and cyberbullies), as well as develops a solution that recognizes the very real constraints placed upon law enforcement officials, such as chain-of-custody issues, and providing proof of integrity of digital evidence.

8.3   Commercial software for monitoring chat

Many commercial products profess to provide parents with the tools to protect their children from Internet predators and cyberbullies. We provide a brief overview of several popular products in this section.

Like most of the parental control products we identified, eBlaster records everything that occurs on a monitored computer and forwards the information to a designated recipient, but does not provide a mechanism for filtering or analyzing all the data it collects (eBlaster 2008). Net Nanny can also record everything, and offers multiple levels of protection for different users (Net Nanny 2008). The latest version of Net Nanny claims to send alerts to parents when it detects predatory or bullying interactions on a monitored computer. The alerts appear to be based on simple keyword matching (PC Mag 2008).

IamBigBrother captures everything on the computer including chats, instant messages, email, and websites (IamBigBrother 2009). The program also records all Facebook and MySpace keystrokes, and captures all passwords typed. IamBigBrother can also take a picture of the screen when certain words are used.

This feature allows parents to identify keywords that they are concerned about (personal information, foul language, sexual terms, etc.). Unfortunately, the program does not include predefined words; parents have to define problematic words themselves (TopTenReviews 2009). The software also captures Internet activity from programs like America Online, MSN, and Outlook Express. The program can record incoming and outgoing Yahoo Mail, Hotmail, and Gmail. IamBigBrother can operate in a stealth mode that cannot be detected by users. Users/children also cannot avoid IamBigBrother by clearing cache or history.

While IamBigBrother appears to focus primarily on keystroke capture and surveillance, Kidswatch Internet Security appears to focus more on blocking (TigerDirect 2009). The program allows parents to control their children's access to inappropriate web content and sends email notifications to parents when their children try to visit blocked or restricted sites. Parents can select content to be restricted from a list of over 60 categories. According to the Kidswatch website: 'Our dynamic content categorization technology attempts to categorize thousands, even millions, of websites based on content.' Parents have the option to override restricted lists if they choose, and are encouraged to submit websites they think should be blocked to the software producer. Kidswatch also supports chat protocols for Yahoo, MSN, ICQ, AIM, and Jabber.

Parents receive email alerts when a 'suspect phrase or word' is encountered in an online chat. The alert report can include the phrase or the entire conversation. The alerts are based on a customizable list of 1630 words and phrases. Although the surveillance and alert features are similar to the one featured in the Net Nanny and IamBigBrother programs, Kidswatch takes this feature one step further by providing information about known sex offenders and on the locations of sex offenders in the user's neighborhood.

Similar to other control programs, the Safe Eyes Parental Control program limits access to restricted sites that fall into 35 predetermined categories of website content (InternetSafety 2009). The program also prevents children from accidentally finding inappropriate sites. When restricted sites are accessed, parents are alerted by email, text message, or phone call.

CyberPatrol provides filtering and monitoring features that can use the company's presets or can be customized by parents (CyberPatrol 2009). Several features that distinguish this program are the ability to customize settings for child, young teen, mature teen, or adult and the ability to block objectionable words and phrases commonly used by cyberbullies and predators. Parents receive weekly and daily reports on web pages visited and length of visits; however, there does not appear to be an alert feature.

Bsecure provides filtering – with 'patent-pending technology and human review' (Bsecure 2009) that blocks offensive websites from users' computers – and reporting options similar to other programs, but this program also offers an Application Control that allows parents to control music sharing, file sharing, and instant messaging programs. The software appears to be similar to CyberPatrol. Bsecure does not offer an alert feature.

The latest versions of Windows Vista and Apple's OS X 10.5 (Leopard) include integrated parental controls. Their features appear to be similar to most commercial monitoring and filtering products and neither operating system, unlike many commercial products, requires an annual subscription (Consumer Search 2008). Unfortunately neither product provides specific protection against predation or cyberbullying.

Finding information about AOL parental controls proved to be fairly difficult without an AOL userid and AOL installed. Like Windows Vista and OS X 10.5, AOL does not require installation of any additional software on the computer being monitored. There is no indication that AOL provides specific features for protection against Internet predators or cyberbullies.

McAfee and Norton are primarily known as antivirus and security software products. Both now offer parental control built in as well. As with the operating system products, the parental controls are designed to block specific websites and monitor online activity in general.

8.4   Conclusions and future directions

The Internet continues to grow and to reach younger audiences. Opportunities for connecting with classmates, friends, and people with shared interests abound. Email, online chat, and social networking sites allow us to interact with people in the same town and people on the other side of the world.

Unfortunately, the opportunity for misuse comes with any new technology. There were sexual predators and bullies long before the advent of the Internet and chat rooms. Cyberbullying and Internet predation threaten minors, particular teens and tweens who do not have adequate supervision when they use the computer. As Internet connectivity moves to the cell phone, the portable gaming device, and the multi-player gaming console, more avenues for contact and exploitation of youth become available.

Our literature review shows that there are few scholars researching cyberpredation and cyberbullying. As more researchers enter this field, future research should attempt to be more proactive in addressing the role that newer technologies, particularly cell phones and peer-to-peer devices, play in new incarnations of cybercrime, like sexting. There is room for researchers in the fields of information retrieval and text mining to contribute solutions to these vexing problems. Classifiers that identify predatory behavior can be developed. New datasets can be collected, labeled, and distributed to other research groups. Collaborations with network engineers, psychologists, sociologists, law enforcement, and communications specialists can provide new insight into understanding, detecting, and stopping cybercrime.

Cybercrime continues to escalate and evolve as new technologies are introduced and as their popularity grows among young people. We have found only three research articles that use text mining techniques to classify cyberpredators and cyberbullies. This interesting and socially relevant subfield of text mining is begging for attention from the research community. The research to date provides a starting point for exploration – an exploration that moves away from solely focusing on the computer platform as the site of cybercrimes to studying the network level as bullying and predation move from text-only, and to include streaming audio and video.

8.5   Acknowledgements

This work was supported in part by the Ursinus College Summer Fellows program. The authors thank Dr Susan Gauch and her students for providing the ChatTrack data, and Dr Nick Pendar for his helpful advice on acquiring the Perverted-justice.com transcripts. We also thank Fundación Barcelona Media (FBM) for compiling and distributing the CAW 2.0 shared task datasets. Our thanks extend to the many students and colleagues in both the Mathematics and Computer Science and Media and Communication Studies Departments at Ursinus College who have provided support and input to this project, as well as to the editors for their patience and feedback.

References

Acar E, Camtepe S, Krishnamoorthy M and Yener B 2005 Modeling and multiway analysis of chatroom tensors. *IEEE International Conference on Intelligence and Security Informatics*.

Agatston P, Kowalski R and Limber S 2007 Students perspectives on cyber bullying. *Journal of Adolescent Health* **41**(6), S59–S60.

Axlerod H and Jay DR 1999 Crime and punishment in cyberspace: Dealing with law enforcement and the courts. *SIGUCCS'99: Proceedings of the 27th Annual ACM SIGUCCS Conference on User Services*, pp. 11–14.

Backstrom L, Huttenlocher D, Kleinberg J and Lan. X 2006 Group formation in large social networks: Membership, growth, and evolution. *Proceedings of the 12th ACM SIGKDD International Conference on Knowledge Discovery and Data Mining, KDD'06*.

Bengel J, Gauch S, Mittur E and R Vijayaraghavan. 2004 ChatTrack: Chat room topic detection using classification. *Second Symposium on Intelligence and Security Informatics*.

Boyd D and Ellison N 2007 Social network sites: Definition, history, and scholarship. *Journal of Computer-Mediated Communication* **13**(1), 210–230.

Bsecure 2009 http://www.bsecure.com/Products/Family.aspx.

Burmester M, Henry P and Kermes LS 2005 Tracking cyberstalkers: A cryptographic approach. *ACM SIGCAS Computers and Society* **35**(3), 2.

Camtepe S, Krishnamoorthy M and Yener B 2004 A tool for Internet chatroom surveillance. *Second Symposium on Intelligence and Security Informatics*.

CAW2.0 2009 http://caw2.barcelonamedia.org/.

Consumer Search 2008 Parental control software review. http://www.consumersearch.com/parental-control-software/review.

Cooke E, Jahanian F and Mcpherson D 2005 The zombie roundup: Understanding, detecting, and disrupting botnets. *Workshop on Steps to Reducing Unwanted Traffic on the Internet (SRUTI)*, pp. 39–44.

CyberPatrol 2009 http://www.cyberpatrol.com/family.asp.

Dewes C, Wichmann A and Feldmann A 2003 An analysis of Internet chat systems. *IMC'03: Proceedings of the 3rd ACM SIGCOMM Conference on Internet Measurement*, pp. 51–64.

eBlaster 2008 http://www.eblaster.com/.

Ellison N, Steinfield C and Lampe C 2007 The benefits of Facebook 'friends': Social capital and college students' use of online social network sites. *Journal of Computer-Mediated Communication* **12**(4), 1143–1168.

FacebookMobile 2009 http://www.facebook.com/mobile/.

Gianvecchio S, Xie M, Wu Z and Wang H 2008 Measurement and classification of humans and bots in internet chat. *SS'08: Proceedings of the 17th Conference on Security Symposium*, pp. 155–169.

Hartigan J and Wong MA 1979 A k-means clustering algorithm. *Applied Statistics* **28**(1), 100–108.

IamBigBrother 2009. http://www.iambigbrother.com/.

ICQ-Sniffer 2009 icq-sniffer.qarchive.org/.

IM MarketShare 2009 http://www.bigblueball.com/forums/general-other-im-news/34413-im-market-share.html/.

Internet Crimes Against Children 2009. http://www.icactraining.org/.

InternetSafety 2009 http://www.internetsafety.com/safe-eyes-parental-control-software.php.

Jones Q, Moldovan M, Raban D and Butler B 2008 Empirical evidence of information overload constraining chat channel community interactions. *Proceedings of the ACM 2008 Conference on Computer Supported Cooperative Work*.

Kontostathis A, Edwards L and Leatherman A 2009 ChatCoder: Toward the tracking and categorization of Internet predators. *Proceedings of the Text Mining Workshop 2009 held in conjunction with the Ninth SIAM International Conference on Data Mining (SDM 2009)*.

Kumar R, Novak J, Raghavan P and Tomkins A 2004 Structure and evolution of blogspace. *Communications of the ACM* **47**(12), 35–39.

Leatherman A 2009 Luring language and virtual victims: Coding cyber-predators' online communicative behavior. Technical report, Ursinus College, Collegeville, PA.

Leskovec J, Lang KJ, Dasgupta A and Mahoney MW 2008 Statistical properties of community structure in large social and information networks *WWW'08: Proceedings of the 17th International Conference on World Wide Web*, pp. 695–704.

Muller M, Raven M, Kogan S, Millen D and Carey K 2003 Introducing chat into business organizations: Toward an instant messaging maturity model. *Proceedings of the 2003 International ACM SIGGROUP Conference on Supporting Group Work*.

Nash KS 2008 A peek inside Facebook. http://www.pcworld.com/business-center/article/150489/a peek inside facebook.html.

National Center for Missing and Exploited Children 2008 http://www.missingkids.com/en US/documents/CyberTiplineFactSheet.pdf.

National Crime Prevention Council 2009a. http://www.ncpc.org/topics/by-audience/cyberbullying/cyberbullying-faq-for-teens.

National Crime Prevention Council 2009b. http://www.ojp.usdoj.gov/cds/internet safety/NCPC/Stop CyberbullyingBeforeItStarts.pdf.

Net Nanny 2008 http://www.netnanny.com/.

Olson L, Daggs J, Ellevold B and Rogers T 2007 Entrapping the innocent: Toward a theory of child sexual predators' luring communication. *Communication Theory* **17**(3), 231–251.

O'Murchu I, Breslin J and Decker S 2004 Online social and business networking communities. Technical report, Digital Enterprise Research Institute (DERI).

PC Mag 2008 Net Nanny 6.0 http://www.pcmag.com/article2/0,2817, 2335485,00.asp.

Pendar N 2007 Toward spotting the pedophile: Telling victim from predator in text chats. *Proceedings of the First IEEE International Conference on Semantic Computing*, pp. 235–241.

Personal Communication 2008 Trooper Paul Iannace, Pennsylvania State Police, Cyber Crimes Division.

Perverted-Justice.com 2008 Perverted justice. www.perverted-justice.com.

Quinlan R 1993 *C4.5: Programs for Machine Learning*. Morgan Kaufmann.

Rawn RWA and Brodbeck DR 2008 Examining the relationship between game type, player disposition and aggression. *Future Play '08: Proceedings of the 2008 Conference on Future Play*, pp. 208–211.

Riffe D, Lacy S and Fico F 1998 *Analyzing Media Messages: Using Quantitative Content Analysis in Research*. Lawrence Erlbaum Associates.

Sipior JC and Ward BT 1999 The dark side of employee email. *Communications of the ACM* **42**(7), 88–95.

TigerDirect 2009 http://www.tigerdirect.com/applications/Search-Tools/item-details.asp?EdpNo=3728335\&CatId=986.

TopTenReviews 2009. http://monitoring-software-review.toptenreviews.com/i-am-big-brother-review.html.

Tuulos V and Tirri H 2004 Combining topic models and social networks for chat data mining. *Proceedings of the 2004 IEEE/WIC/ACM International Conference on Web Intelligence*, pp. 235–241.

Van Dyke N, Lieberman H and Maes P 1999 Butterfly: A conversation-finding agent for Internet relay chat. *Proceedings of the 4th International Conference on Intelligent User Interfaces*.

Williams K and Guerra N 2007 Prevalence and predictors of Internet bullying. *Journal of Adolescent Health* **41**(6), S14–S21.

Witten I and Frank E 2005 *Data Mining: Practical Machine Learning Tools and Techniques*. Morgan Kaufmann.

Yin D, Xue Z, Hong L, Davison BD, Kontostathis A and Edwards L 2009 Detection of harassment on Web 2.0. *Proceedings of the Content Analysis in the Web 2.0 (CAW2.0) Workshop at WWW2009*.

Part III

TEXT STREAMS

9

Events and trends in text streams

Dave Engel, Paul Whitney and Nick Cramer

9.1 Introduction

Text streams – collections of documents or messages that are generated and observed over time – are ubiquitous. Our research and development are targeted at developing algorithms to find and characterize changes in topic within text streams. To date, this research has demonstrated the ability to detect and describe (1) short-duration atypical events and (2) the emergence of longer term shifts in topical content. This technology has been applied to predefined temporally ordered document collections but is suitable also for application to near-real-time textual data streams.

Massive amounts of text stream data exist and are readily available, especially over the Internet. Analyzing this text data for content and for detecting change in topic or sentiment can be a daunting task. Mathematical and statistical methods in the area of data mining can be very helpful to the analyst looking for these changes. Specifically, we have implemented some of these techniques into a *surprise* event and *emerging* trend detection technology designed to monitor a stream of text or messages for changes within the content of that data stream.

Some of the event types that one might want to detect in a text stream (which could be a sequence of news articles, a sequence of messages, or an evolving dialogue) are shown in Figure 9.1. In each case, time is along the x-axis. The y-axis corresponds to some measure of topic (such as the number of words or events

Text Mining: Applications and Theory   edited by Michael W. Berry and Jacob Kogan
© 2010, John Wiley & Sons, Ltd

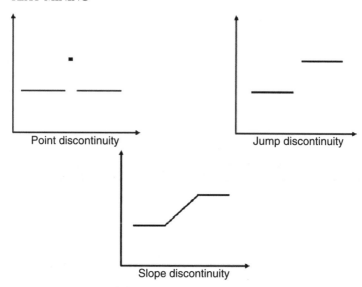

Figure 9.1   Typical event or trend types.

that occur within the data). In the context of a text stream, a *point discontinuity* in topics could correspond to a single time step with a relatively unique content. A *jump discontinuity* could correspond to an abrupt change in the content of the text stream. A *slope discontinuity* could correspond to a ramping up (or down) in a topic for that text stream.

Typically, jump and point discontinuities are detected more readily than slope discontinuities (Eubank and Whitney et al. 1989). For our terminology, we refer to the instantaneous discontinuity types (point or jump) as a *surprise* event (see Grabo (2004) for more information on *surprise* events). We define an *emerging* trend as a change in topic for an extended period of time, as illustrated by the *jump discontinuity* or the *slope discontinuity* (see Kontostathis et al. (2003) for a more concise definition of *emerging* trend).

Much of the research in information mining from text streams focuses either on describing new events and salient features or in clustering documents (He et al. 2007; Kumaran and Allan 2004; Mei and Zhai 2005). For instance, the goal of the Topic Detection and Tracking (TDT) Research Program (Allan 2002) was to break down the text into individual news stories, to monitor the stories for events that have not been seen before, and to gather the stories into groups that each discuss a single topic. This program used a training set to identify stories (topics) to track. A good source of research in trend analysis was compiled in *Survey of Text Mining: Clustering, Classification, and Retrieval* (Kontostathis et al. 2003) and also in the article 'Detecting emerging trends from scientific corpora' (Le et al. 2005). In both, the main focus is tracking defined topics and trying to detect changes.

The difference in our approach is that we monitor and evaluate the occurrence of individual terms (the least common denominator between documents) for changes over time. Once individual terms have been determined as *surprising* or *emerging*, then terms related temporally are identified to help the analyst identify the story/topic involved with the *surprising (emerging)* terms. As a preprocessing step, a text analysis tool is used to extract words from the text stream and give information about terms within the documents. With this information, mathematical algorithms are used to score each term. Using these scores (statistical metrics, which we call *surprise* or *emergence* statistics), we evaluate each term over the period represented by the text stream. When a sufficiently *surprising (emerging)* term occurs, related terms (based on the temporal profile) are found and are useful in explaining the broader nature of the event.

Detected events and the explanatory terms can be represented in a variety of ways. From our experience, graphical representations tend to be the most desirable (if not most useful) form for the analysts.

A description of the data (text streams) and the extraction and reduction of relevant features are discussed in the next two sections. The methodology for the detection of *(surprising)* events and *(emerging)* trends is discussed in Sections 9.4 and 9.5. In Section 9.6, we discuss temporally related terms and present an example to illustrate the capabilities of our technology. The last two sections discuss differences in our algorithms, contrast our algorithms with other topicality measures, and summarize our technology development.

9.2   Text streams

Many text analysis tools operate on a fixed collection of text documents. For certain tasks, a fixed text collection is appropriate. However, information analysis professionals often seek to discover and track *surprising* events and *emerging* trends over time and in a timely fashion. A text stream is necessary to support this analytic task (Hetzler et al. 2005). Text streams are often rich with *surprising* or *emerging* events and interesting topic evolution over time. Detection of these events can provide information analysts with valuable information and clues about their content.

For our methodology, a document is simply defined as a unique collection of text. A text stream is a collection of documents in which each document has an associated time stamp. Each document typically contains metadata describing the publication time and date or is assigned a time and date when collected. In either case, the time stamp allows us to orient the document text in the temporal stream.

Text streams are generated from a variety of data sources. Some examples include journal publications, conference abstracts, really simple syndication (RSS) news feeds, blog postings, and email transmissions. To handle the collection of text streams from the variety of sources, we have developed and implemented resources. These resources include:

- conference PDF text extractors;

- Outlook email harvesters;

- RSS news feed harvesters;

- blog post harvesters.

An information analyst might want to follow information only within a window of time. Text streams can evolve over time, with not only new content being added to the collection but also old content being removed. The resource that we have implemented supports an evolving text collection which helps an analyst focus on the most relevant and timely information.

9.3   Feature extraction and data reduction

Once the data (text stream) has been collected, the next step involves processing the data (documents) to evaluate suitability of the data content and prepare the data for subsequent processing. We use IN-SPIRE for this processing (IN-SPIRE 2009). IN-SPIRE is a text analysis and visualization tool that statistically analyzes unstructured text within a collection of documents, identifies topics (i.e. terms with high-frequency and nonuniform distributions), and visually clusters the documents based on their topical similarity. IN-SPIRE provides the following capabilities for the preprocessing steps for event and trend detection:

1. *Dataset evaluation.* An initial evaluation of the document collection is performed by information analysts to determine if the datasets are sufficiently rich.

2. *Content identification and index creation.* Relevant content is extended from text, largely ignoring many of the other categorical fields (e.g. authors and place names).

3. *Topical feature selection.* Vocabulary terms that are statistically good discriminators are identified. In addition, relevant terms/keywords to the domain can be provided to augment and enrich the automatic topic selection process. This topical term and phrase identification process acts as a dimensionality reduction that helps focus the analyses.

In our modeling, the (document frequency) temporal profile for each term is the dependent variable. Therefore, the selection of terms that represent the document set is a key task. This task is accomplished within IN-SPIRE. An important feature of this capability is the automatic keyword extraction. This capability allows keywords to be single words or phrases that reflect the content of a document. An in-depth description of this technology is included in the first chapter of this book.

9.4   Event detection

Our research is focused on processing massive amounts of text streams to identify events that have just occurred or are currently occurring. You can think of this as a possible triage capability that an analyst needs to identify (*surprising*) events so that he or she can delve into the material to gain in-depth insight. However, finding these events in a timely fashion is not an easy task.

Different algorithms for detecting *surprising* events have been researched and five of these algorithms have been implemented into our research toolkit. For each algorithm, the unit of calculation is a term or keyword that can be a single word or multiple words. Each of our algorithms requires a preprocessing of the time-sequenced documents (as described in Section 9.3).

In statistics, we deal with numbers. Therefore, the first step in analyzing text using statistical models (algorithms) is to convert the text to numbers. For our analytical methods, we have done this by counting the number of documents that contain a given term (keyword). For each document, a time stamp is identified, allowing our analysis to be done temporally. The overall time interval for which the documents occur is divided into equally spaced time bins (we may use hourly, daily, or even weekly intervals, depending on the temporal granularity of the data being examined). The number of documents that contain a specific term within each time bin becomes the main variable of our analysis (call it a temporal profile).

We analyze each temporal profile (one for each term) using one of our algorithms and define a *surprise* statistic, which is calculated in each time interval. Figure 9.2 illustrates the temporal profiles used in our analysis. Seven profiles are shown; each profile represents the number of documents that contain the specific term within each time bin. Each term is normalized individually (by the maximum number of occurrences of the individual term) and then plotted (i.e. the vertical axis for each term is scaled 0 to 1). The maximum number of occurrences of each individual term within a single time interval is shown on the right side of each profile (e.g. six documents for the term *influenza*).

Also illustrated in Figure 9.2 is our *surprise* text mining methodology (Whitney et al. 2009). For this method, the number of occurrences (x_i) within a single time step/bin is compared against the number of occurrences within a previous time window (multiple consecutive time bins). The comparison is repeated for every time step (i.e. moving time window). The time bin with the maximum *surprise* score is considered the location of the *surprise* event. These maximum values will be identified by the circles for each term. The previous time window for the location of the *surprise* event starts at the vertical line and ends at the time bin represented by the circle (but not including this time step).

The task is then to compare the document counts (number of documents containing a specific term), at a single time step (step i), to the document counts in the time window just before the current time step (np consecutive time steps/bins).

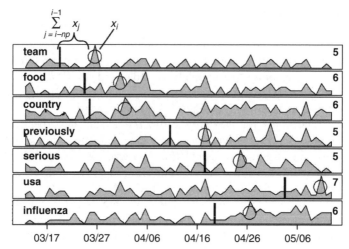

Figure 9.2   Modeling scheme and temporal profiles for the event detection algorithms (term label on left side of each profile and maximum number of occurrences per profile on right side of each profile).

The goal is to find the times when these two measurements (counts) are not (statistically) the same. Think of this like a hypothesis test in statistics: we define the null hypothesis (H_o) and alternate hypothesis (H_a) as

$$H_o: x_i = \frac{1}{np} \sum_{j=i-np}^{i-1} x_j, \text{ and}$$

$$H_a: x_i \neq \frac{1}{np} \sum_{j=i-np}^{i-1} x_j.$$

The goal of a hypothesis test is to reject the null hypothesis and accept the alternate hypothesis. We have developed our algorithms with this in mind. The first *surprise* algorithm is based on a chi-square statistic (Pearson method) constructed from the following 2×2 table (Agresti 2002):

$$\begin{pmatrix} x_i & N_i - x_i \\ \sum_{j=i-np}^{i-1} x_j & \sum_{j=i-np}^{i-1} N_j - \sum_{j=i-np}^{i-1} x_j \end{pmatrix}$$

where, for this table, x_i is the count (number of documents containing a specific term) at the ith time step/bin, N_i is the total number of documents at the ith time step, $\sum x_j$ is the sum of the document counts containing the term in the (np) time steps prior to the ith time step, and $\sum N_j$ is the total number of documents in the (np) time steps prior to time t (time at the ith time step). The amount of time (both the width of a time interval and the number of time windows) is a user-selected parameter of the procedure. A value sufficiently large for a chi-square statistic is one way to flag a *surprising* event/term. This statistic looks for deviations in the number of occurrences of a specific term normalized by the total number of documents (within the same time interval).

The formula used for the chi-square statistic is

$$\chi^2 = \frac{n_{..}\left(|n_{11}n_{22} - n_{12}n_{21}| - \frac{1}{2}Yn_{..}\right)^2}{n_{1.}n_{2.}n_{.1}n_{.2}}, \tag{9.1}$$

where the previous 2×2 frequency table is rewritten as

$$\begin{pmatrix} n_{11} & n_{12} \\ n_{21} & n_{22} \end{pmatrix}$$

and

$$n_{1.} = n_{11} + n_{12},$$

$$n_{2.} = n_{21} + n_{22},$$

$$n_{.1} = n_{11} + n_{21},$$

$$n_{.2} = n_{12} + n_{22}, \text{ and}$$

$$n_{..} = n_{11} + n_{12} + n_{21} + n_{22}.$$

Also, Y in Equation (9.1) is either 0 or 1. If Y is 1, the Yates continuity correction is applied for the low sample size in which the count in at least one cell is ≤ 5 (Fleiss 1981).

The second algorithm for calculating the *surprise* statistic is another form of the chi-square algorithm known as the likelihood ratio. The likelihood ratio (for a hypothesis) is the ratio of the maximum value of the likelihood function over the subspace represented by the hypothesis, to the maximum value of the likelihood function over the entire parameter space (Dunning 1993). This statistic is calculated using the same 2×2 table as above and is as follows:

$$\chi^2 = \frac{1}{2}\left(n_{11}\log\frac{n_{11}}{m_{11}} + n_{12}\log\frac{n_{12}}{m_{12}} + n_{21}\log\frac{n_{21}}{m_{21}} + n_{22}\log\frac{n_{22}}{m_{22}}\right), \tag{9.2}$$

where

$$m_{11} = (n_{11} + n_{12})(n_{11} + n_{21}),$$

$$m_{12} = (n_{11} + n_{12})(n_{12} + n_{22}),$$

$$m_{21} = (n_{11} + n_{21})(n_{21} + n_{22}), \text{ and}$$

$$m_{22} = (n_{12} + n_{22})(n_{21} + n_{22})$$

Another of our algorithms for calculating the *surprise* statistic is a Gaussian algorithm. The Gaussian statistic is based on comparing the observed value x_i to the average over the previous values $(1/np)\sum x_j$, normalized by the standard deviation of these previous values. We put a floor of 1.0 on the standard deviation because we are dealing with count data. This statistic is

$$G = \frac{x_i - \frac{1}{np}\sum_{j=i-np}^{i-1} x_j}{s \cdot \left(1 + \frac{1}{np}\right)}, \tag{9.3}$$

where np is the number of time intervals in the previous time windows and s is the standard deviation.

Finally, combining the previous algorithms (chi-square and Gaussian) forms the final two algorithms within our toolkit for the *surprise* statistic. Each combined statistic is accomplished by taking the square root of the chi-square statistic plus the absolute value of the Gaussian statistic, as follows:

$$C_{surprise} = \sqrt{\chi^2} + |G|. \tag{9.4}$$

9.5 Trend detection

Starting from the algorithms for detecting *surprising* events, we developed a modeling scheme for detecting (*emerging*) trends. The modeling scheme is shown in Figure 9.3. In this figure, x is the number of documents within a time step that contains the specific term, i is the current time step (time interval/bin), np is the number of time steps in the previous time window, and nc is the number of time steps in the current time window (Engel et al. 2009).

For detecting trends, we compare the document counts (number of documents containing a specific term) of the current time window (current time step i plus the next (nc) time steps) to the document counts in the time window just prior to the current time step (np consecutive time steps). The goal is to find the times when these two measurements (counts) are not (statistically) the same. Similar to the *surprise* statistic, we calculate an *emergence* statistic in which we define the null hypothesis (H_o) and alternate hypothesis (H_a) as

$$H_o: \frac{1}{nc}\sum_{j=i}^{i+nc} x_j = \frac{1}{np}\sum_{j=i-np}^{i-1} x_j, \text{ and}$$

$$H_a: \frac{1}{nc}\sum_{j=i}^{i+nc} x_j > \frac{1}{np}\sum_{j=i-np}^{i-1} x_j.$$

The event detection technology is designed to be used to monitor a stream of text or messages for changes within the *content* of that stream. An analyst might

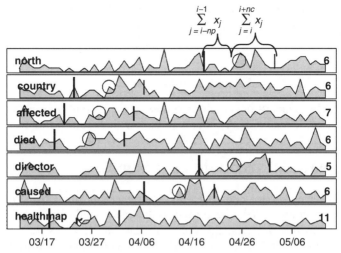

Figure 9.3   Modeling scheme and temporal profiles for the trend detection algorithms.

be watching a news feed or exploring a large collection of message traffic. This technology would be used to detect and describe changes in those text streams.

For the *emergence* statistic, the two chi-square algorithms are the same as the algorithms for the *surprise* statistic (Equations (9.1) and (9.2)), but the 2×2 frequency table is replaced by

$$
\begin{pmatrix}
\sum\limits_{j=i}^{i+nc} x_j & \sum\limits_{j=i}^{i+nc} N_j - \sum\limits_{j=i}^{i+nc} x_j \\
\sum\limits_{j=i-np}^{i-1} x_j & \sum\limits_{j=i-np}^{i-1} N_j - \sum\limits_{j=i-np}^{i-1} x_j
\end{pmatrix}
$$

where, for this table, $\sum x_j$ in the first row is the sum of all the documents containing the individual term in the current time window, $\sum N_j$ in the first row is the total number of documents within this time period, $\sum x_j$ in the second row is the sum of all the documents containing the term within the period prior to the current time step (previous window), and $\sum N_j$ in the second row is the total number of documents within this (previous) time window. The number of time steps within each interval (previous window and current window) is a user-selected parameter of the procedure. (Note that these window sizes need not be equal.)

For detecting trends, the Gaussian algorithm is modified from the *surprise* implementation to incorporate the multiple time steps in the current time window (time past the current time step, i). The new Gaussian algorithm is defined by

$$G = \frac{\frac{1}{nc}\sum_{j=i}^{i+nc} x_j - \frac{1}{np}\sum_{j=i-np}^{i-1} x_j}{\sqrt{\frac{s_i}{nc} + \frac{s_j}{np}}},$$

where s_i is the standard deviation of counts in the current time window and s_j is the standard deviation of the counts in the previous time window.

9.6   Event and trend descriptions

To illustrate the (*surprise*) event detection and (*emerging*) trend detection capabilities, both technologies have been used in the analysis illustrated in Figures 9.4 through 9.8. In this analysis, the source of the data (text) is the International Society for Infectious Diseases (ProMED-mail 2009). This website is a global electronic reporting system for outbreaks of emerging infectious diseases and toxins, open to all sources. Contributions to this site tend to be from medical professions. In Figure 9.4, the documents from this (ProMed-mail) dataset have been cumulated into one-day time intervals, with the number of documents per time interval displayed.

The results from both the *surprise* analysis and the *emergence* analysis are shown in Figures 9.5 through 9.8. Figures 9.5 and 9.6 show results from the

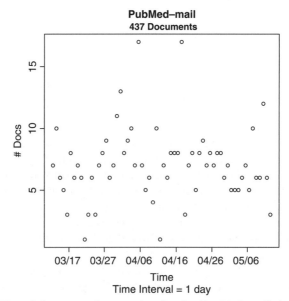

Figure 9.4   Binned document frequencies for the ProMed-mail dataset, one-day time resolution.

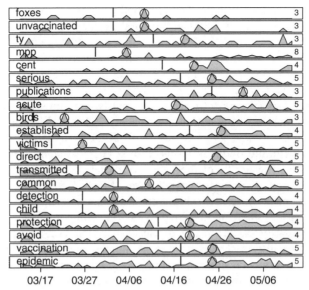

temporal profiles, max Surprise, bin.width = 1 day, # bins = 7, PubMed–mail dataset

Figure 9.5   Temporal profiles sorted by the chi-square (Pearson) surprise *score (ProMed-mail).*

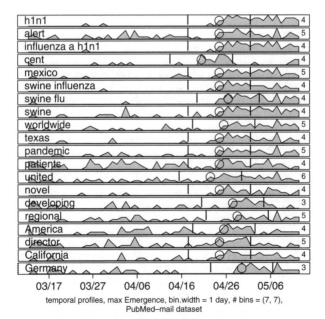

temporal profiles, max Emergence, bin.width = 1 day, # bins = (7, 7),
PubMed–mail dataset

Figure 9.6   Temporal profiles sorted by the chi-square (Pearson) emergence *score (ProMed-mail).*

chi-square (Pearson) algorithms. The temporal profiles for the top 20 *surprising* terms are shown in Figure 9.5. The temporal profiles for the top 20 *emerging* terms are shown in Figure 9.6. From these two plots, the main topic within this dataset becomes obvious (H1N1, the swine flu outbreak of 2009). On April 24, the *surprise* analysis (Figure 9.5) starts to select terms that first appear about the swine flu outbreak (*serious*, *vaccination*, *epidemic*). However, the results of the *emergence* analysis (Figure 9.6) clearly explain when and what occurred. The results of using the Gaussian algorithms to analyze this ProMed-mail dataset are shown in Figures 9.7 and 9.8. The results from the Gaussian *surprise* analysis show that no swine flu outbreak terms were selected as significantly *surprising* for this analysis. The results of the *emergence* analysis, however, did show the selection of several (swine flu) relevant terms (Figure 9.8).

Similarities between terms within a given set can give an analyst more information than just a single term can provide (including multi-term keywords). We assess similarity based on the distances between vectors of the temporal occurrence of each term. There are a large number of candidate algorithms for calculating distances between temporal profiles. Our preferred implementation is based on the correlation function between the vectors and is, for two such vectors (x, y), equal to $1 - |corr(x, y)|$. This distance often results in interpretable term groupings (Kaufman and Rousseeuw 1990). Using combined related term profiles, one can gain more detailed information about the events. For illustration, Figure 9.9 shows the related terms for the term *mexico* (from the analysis of the

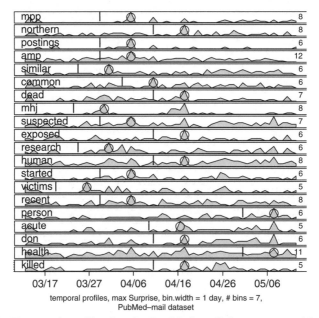

Figure 9.7   Temporal profiles for the ProMed-mail dataset, sorted by the Gaussian surprise *score.*

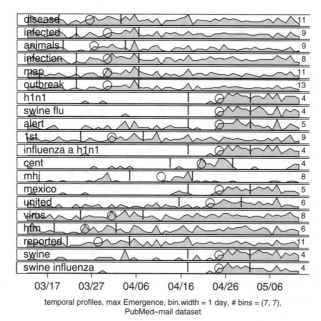

Figure 9.8   Temporal profiles for the ProMed-mail dataset, sorted by the Gaussian emergence *score.*

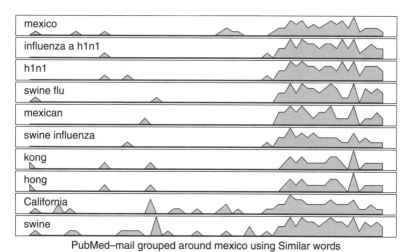

Figure 9.9   Temporal profiles for the term mexico *and the top nine related terms (ProMed-mail dataset).*

ProMed-mail dataset). From this, it is obvious that the main topic about this term (*mexico*) is the 2009 swine flu (H1N1) outbreak.

9.7 Discussion

In the previous section, the *surprise* and *emergence* algorithms were used to analyze the ProMed-mail dataset. From Figure 9.4, we see that the maximum number of documents (reports) for a single day (from March 13 through May 13) was 17. In Figure 9.6, we see that the maximum number of documents that contained the term *h1n1* was only 4 (number on the right hand side of each temporal profile). Because of the low number of term occurrences and document counts, the *surprise* algorithms did not produce the desired results compared to the results from the *emergence* algorithms.

A comparison of the *surprise* statistic (maximum value for each term) and the *emergence* statistic is shown in Figure 9.10. Also shown in this figure is a comparison of the IN-SPIRE *topicality* score to the *surprise* and *emergence* statistic. The IN-SPIRE *topicality* score is a measure that defines discriminating terms within a set of documents. This comparison was done using the ProMed-mail dataset and the chi-square (Pearson) algorithms. The fundamental observation is that the metrics are uncorrelated, at least for this corpora, because no correlation is seen in any of these plots (or very low correlation for the surprise–emergence

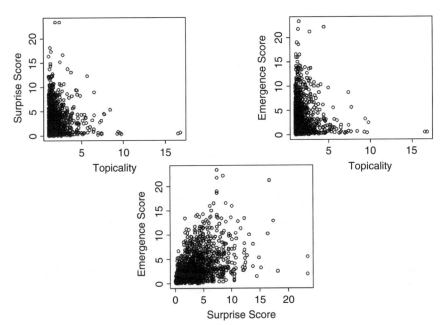

Figure 9.10 Comparison of topicality (topicality), event detection (surprise score), and trend detection (emergence score) algorithms (ProMed-mail dataset).

plot), which suggests that these three statistics provide different information about the dataset.

9.8   Summary

Mathematical and statistical methods in the area of text mining can be very helpful for the analysis of the massive amounts of text stream data that exists. Analyzing this data for content and for detecting change can be a daunting task. Therefore, we have implemented some of these text mining techniques into a *surprise* event and *emerging* trend detection technology that is designed to monitor a stream of text or messages for changes within the content of that data stream.

In this chapter, we have described our algorithmic development in the area of detecting evolving content in text streams (events and trends). We have compared our results to text analysis results on a static document collection and found that our techniques produce results that are different and enhance those results.

A recent dataset was analyzed using our *surprise* and *emergence* algorithms. In this analysis, the *emergence* algorithms did a very good job of finding the *emergence* of the most relevant subject matter (H1N1, swine flu outbreak) and when the event began (April 24, 2009).

To help understand the important topics defined by each term (keyword), related terms are found. For the swine flu analysis, the term *mexico* was found to be a significant *emerging* term. The related term analysis showed that this term was temporally related to the swine flu (H1N1) outbreak (2009).

9.9   Acknowledgements

The authors of this chapter would like to thank Andrea Currie for her editorial review and Guang Lin for his LaTeX help and expertise.

References

Agresti A 2002 *Categorical Data Analysis* 2nd edn. John Wiley & Sons, Inc.

Allan J 2002 *Topic Detection and Tracking: Event-based Information Organization*. Kluwer Academic.

Dunning T 1993 Accurate methods for the statistics of surprise and coincidence. *Computational Linguistics* **19**(1), 61–74.

Engel D, Whitney P, Calapristi A and Brockman F 2009 Mining for emerging technologies within text streams and documents. *Ninth SIAM International Conference on Data Mining*. Society for Industrial and Applied Mathematics.

Eubank R and Whitney P 1989 Convergence rates for estimation in certain partially linear models. *Journal of Statistical Planning and Inference* **23**, 33–43.

Fleiss J 1981 *Statistical Methods for Rates and Proportions* 2nd edn. John Wiley & Sons, Inc.

Grabo C 2004 *Anticipating Surprise: Analysis for Strategic Warning*. University Press of America.

He Q, Chang K, Lim E and Zhang J 2007 Bursty feature representation for clustering text streams. *Seventh SIAM International Conference on Data Mining*, pp. 491–496. Society for Industrial and Applied Mathematics.

Hetzler E, Crow V, Payne D and Turner A 2005 Turning the bucket of text into a pipe. *IEEE Symposium on Information Visualization*, pp. 89–94.

IN-SPIRE 2009 http://in-spire.pnl.gov *Pacific Northwest National Laboratory*.

Kaufman L and Rousseeuw P 1990 *Finding Groups in Data: An Introduction to Cluster Analysis*. John Wiley & Sons, Inc.

Kontostathis A, Galitsky L, Pottenger W, Roy S and Phelps D 2003 A survey of emerging trend detection in textual data mining. in: *Survey of Text Mining: Clustering, Classification, and Retrieval*. Springer.

Kumaran G and Allan J 2004 Text classification and named entities for new event detection. *ACM SIGIR Conference* pp. 297–304.

Le M, Ho T and Nakamori Y 2005 Detecting emerging trends from scientific corpora. *ACM SIGIR Conference* pp. 45–50.

Mei Q and Zhai C 2005 Discovering evolutionary theme patterns from text: An exploration of temporal text mining. *KDD, 11th ACM SIGKDD International Conference on Knowledge Discovery and Data Mining*, pp. 198–207.

ProMED-mail 2009 http://www.promedmail.org.

Whitney P, Engel D and Cramer N 2009 Mining for surprise events within text streams. *Ninth SIAM International Conference on Data Mining*, pp. 617–627. Society for Industrial and Applied Mathematics.

10

Embedding semantics in LDA topic models

Loulwah AlSumait, Pu Wang, Carlotta Domeniconi and Daniel Barbará

10.1 Introduction

The huge advancement in databases and the explosion of the Internet, intranets, and digital libraries have resulted in giant text databases. It is estimated that approximately 85% of worldwide data is held in unstructured formats with an increasing rate of roughly 7 million digital pages per day (White 2005). Such huge document collections hold useful yet implicit and nontrivial knowledge about the domain. Text mining (TM) is an integral part of data mining that is aimed at automatically extracting such knowledge from the unstructured textual data. The main tasks of TM include text classification, text summarization, document and/or word clustering, in addition to classical natural language processing tasks such as machine translation and question-answering. The learning tasks are more complex when processing text documents that arrive in discrete or continuous streams over time.

Topic modeling is a newly emerging approach to analyze large volumes of unlabeled text (Steyvers and Griffiths 2005). It specifies a statistical sampling technique to describe how words in documents are generated based on (a small set of) hidden topics. In this chapter, we investigate the role of prior knowledge semantics in estimating the topical structure of large text data in both batch and online modes under the framework of latent Dirichlet alglocation (LDA) topic

Text Mining: Applications and Theory   edited by Michael W. Berry and Jacob Kogan
© 2010, John Wiley & Sons, Ltd

modeling (Blei et al. 2003). The objective is to enhance the descriptive and/or predictive model of the data's thematic structure based on the embedded prior knowledge about the domain's semantics.

The prior knowledge can be either external semantics from prior-knowledge sources, such as ontologies and large universal datasets, or a data-driven semantics which is a domain knowledge that is extracted from the data itself. This chapter investigates the role of semantic embedding in two main directions. The first is to embed semantics from an external prior-knowledge source to enhance the generative process of the model parameters. The second direction which suits the online knowledge discovery problem is to embed data-driven semantics. The idea is to construct the current LDA model based on information propagated from topic models that were learned from previously seen documents of the domain.

10.2 Background

Given the unstructured nature of text databases, many challenges face TM algorithms. First, there are a very high number of possible features to represent a document. Such features can be derived from all the words and/or phrase types in the language. Furthermore, in order to unify the data structure of documents, it is necessary to use a dictionary of all the words to represent a document, which results in a very sparse representation. Another critical challenge stems from the complex relationships between concepts and from the ambiguity and context sensitivity of words in text. Thus, a good TM algorithm must be efficient to process such large and challenging data so that the documents are represented in short descriptions in which only the essential and most discriminative information is preserved. The rest of this section is focused on three major advancements to solve this problem, then the LDA topic models will be introduced in Section 10.3.

10.2.1 Vector space modeling

The first major progress in text processing was due to the vector space model (Salton 1983), in which a document is represented as a vector of dimension W, $\mathbf{w}_d = (w_{1d}, \ldots, w_{Wd})$, where each dimension is associated with one term of the dictionary. Each entry w_{id} is the *term frequency – inverse document frequency* (tf-idf) of the term i in document d given by $w_{id} = n_{id} \times \log(D/n_i)$. The local frequency of the term (n_{id}) is weighted by its global frequency in the whole corpus to reduce the importance of common words that appear in many documents since they are naturally bad discriminators. To represent the whole corpus, the term – document matrix, X, is constructed. X is a $W \times D$ matrix whose rows are indexed by the terms of the dictionary and whose columns are indexed by the documents.

Although the VSM has empirically shown its effectiveness and is widely used, it suffers from a number of inherent shortages to capture inter- and intra-document statistical structure and provides a small reduction only in the description of the corpus.

10.2.2   Latent semantic analysis

To address the shortages of the VSM, researchers in information retrieval (IR) have introduced latent semantic analysis (LSA) (Deerwester et al. 1990), which is a factor analysis that reduces the term – document matrix to a K-dimensional subspace that captures most of the variance in the corpus. By computing the singular value decomposition (SVD), the term – document matrix X is decomposed into three matrices $X = U \Sigma V^T$. The rows in U give the occurrence of the original words which correspond to the K *concepts* of the new factor space, while the columns in V give the relation between the documents and each of the K concepts.

Although LSA overcomes some of the drawbacks of the VSM, it suffers from a number of limitations. First, given the high-dimensionality nature of text data, computation of the SVD is expensive. In addition, the new feature space is very difficult to interpret since each dimension is a linear combination of a set of words from the original space. LSA is also not generalizable to incorporate other side information such as time and author.

10.2.3   Probabilistic latent semantic analysis

Researchers have proposed statistical approaches to understand LSA, some of whom have discussed its relationship to Bayesian methods (Story 1996) and generative probabilistic models (Papadimitriou et al. 2000). As a major advance in the application of Bayesian methods to document modeling, Hofmann (1999) introduced probabilistic latent semantic analysis (pLSA), also called the *aspect model*, as an alternative to LSA. It is a latent variable model that associates an unobserved class (aspect) variable z_k with each document d and represents each aspect by a distribution over words $p(\mathbf{w}|\mathbf{z})$. The pLSA model is parameterized by the joint distribution of a document d and a word w_{di} that appears in it, $p(d, w_{di}) = p(d) \sum_{z=1}^{K} p(w_{di}|z)p(z|d)$.

A graphical model of pLSA is shown in Figure 10.1. Given the hidden aspects, the documents and words are conditionally independent. In addition, pLSA allows the documents to be associated with a mixture of topics weighted by the posterior $p(\mathbf{z}|d)$.

The generative process of a model specifies a probabilistic sampling procedure that describe how words in documents can be generated based on the hidden topics. Thus, the generative process of the pLSA is as follows:

1. Draw a document with probability $p(d)$.

2. For each word i in document d:

    (a) Draw a latent aspect z_i with probability $p(z_i|d)$.

    (b) Draw a word w_{di} with probability $p(w_{di}|z_i)$.

Nonetheless, this is not a true generative model as the variable d is a dummy random variable that is indexed by the documents in a training set (Blei et al.

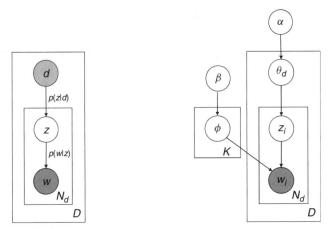

Figure 10.1   A graphical model of pLSA (left) and LDA (right).

2003). As a consequence, pLSA is inclined to overfit the training data, which harms its ability to generalize the inferred aspect model to generate previously unseen documents.

Despite its limitation, pLSA has influenced a huge amount of work in statistical machine learning and TM. As a result, a class of statistical models, named probabilistic topic models (PTMs), have been created to uncover the underlying structure of large collections of discrete data, such as text. PTMs are generative models of documents that assume the existence of hidden variables, representing topics associated with the observed text documents which are responsible for the patterns of word use. Topic models are aimed at discovering these hidden variables based on hierarchical Bayesian analysis. Among the variety of topic models proposed, LDA (Blei et al. 2003) is a truly generative model that is capable of generalizing the topic distributions so that it can be used to generate unseen documents as well.

10.3   Latent Dirichlet allocation

The LDA PTM is a three-level hierarchical Bayesian network that represents the generative probabilistic model of a corpus of documents. The basic idea is that documents are represented by a mixture of topics where each topic is a latent multinomial variable characterized by a distribution over a fixed vocabulary of words. The completeness of the LDA's generative process for documents is achieved by considering Dirichlet priors on the document distributions over topics and on the topic distributions over words. This emerging approach has been successfully applied to find useful structures in many kinds of documents, including emails, the scientific literature (Griffiths and Steyvers 2004), libraries of digital books (Mimno and McCallum 2007), and news archives (Wei and Croft 2006).

This section introduces the LDA topic model with a brief description of its graphical model and generative process (Section 10.3.1) and the posterior inference (Section 10.3.2). The section concludes with a brief review of an online version of LDA, namely OLDA.

10.3.1 Graphical model and generative process

LDA relates words and documents through latent topics based on the bag-of-words assumption, i.e. the *exchangeability*, for the words in a document and for the documents in a corpus. The graphical model of LDA is given in Figure 10.1. The documents θ are not directly linked to the words **w**. Rather, this relationship is governed by additional latent variables, z, introduced to represent the responsibility of a particular topic in using that word in the document, i.e. the topic(s) that the document is focused on. By introducing the Dirichlet priors α and β over the document and topic distributions, respectively, the generative model of LDA is complete and is capable of processing unseen documents.

So, the structure of the LDA model allows the interaction of the observed words in documents with structured distributions of a *hidden variable model* (Blei et al. 2003). Learning the structure of the hidden variable model can be achieved by inferring the posterior probability distribution of the hidden variables, i.e. the topical structure of the collection, given the observed documents. This interaction can be viewed in the generative process of LDA:

1. Draw K multinomials ϕ_k from a Dirichlet prior β, one for each topic k.

2. Draw D multinomials θ_d from a Dirichlet prior α, one for each document d.

3. For each document d in the corpus, and for each word w_{di} in the document:

   (a) Draw a topic z_i from multinomial θ_d; $(p(z_i|\alpha))$.

   (b) Draw a word w_i from multinomial ϕ_z; $(p(w_i|z_i, \beta))$.

Inverting the generative process, i.e. fitting the hidden variable model to the observed data (words in documents), corresponds to inferring the latent variables and, hence, learning the distributions of underlying topics. The hidden structure of topics in the LDA model is described by the posterior distribution of the hidden variables given the D documents

$$p(\Theta, \mathbf{z}, \Phi|\mathbf{w}, \alpha, \beta) = \frac{p(\mathbf{w}, \Theta, \mathbf{z}, \Phi|\alpha, \beta)}{\int_{\phi_{1:K}} \int_{\theta_{1:D}} p(\mathbf{w}|\alpha, \beta)}. \tag{10.1}$$

10.3.2 Posterior inference

In LDA, exploring the data and extracting the topics correspond to computing the posterior expectations. These are the topic probability over terms ($E(\Phi|\mathbf{w})$), the document proportions over topics ($E(\Theta|\mathbf{w})$), and the topic assignments of words ($E(\mathbf{z}|\mathbf{w})$). Although the LDA model is relatively simple, exact inference of the posterior distribution in Equation (10.1) is intractable (Blei et al. 2003). The solution is to use sophisticated approximations such as variational expectation maximization (Blei et al. 2003) and expectation propagation (Minka and Lafferty 2002).

Griffiths and Steyvers (2004) proposed a simple and effective strategy for estimating ϕ and θ. It is an approximate iterative technique that is a special form of Markov chain Monte Carlo (MCMC) methods. Gibbs sampling is able to simulate a high-dimensional probability distribution $p(\mathbf{x})$ by iteratively sampling one dimension x_i at a time, conditioned on the values of all other dimensions, which is usually denoted $\mathbf{x}_{\neg i}$.

Under Gibbs sampling, ϕ and θ are not explicitly estimated. Instead, the posterior distribution over the assignments of words to topics, $P(\mathbf{z}|\mathbf{w})$, is approximated by means of the Monte Carlo algorithm, see Heinrich (2005) for a detailed derivation of the algorithm. Gibbs sampling iterates over each word token in the text collection in a random order and estimates the probability of assigning the current word token to each topic ($P(z_i = j)$), conditioned on the topic assignments to all other word tokens ($\mathbf{z}_{\neg i}$) as (Griffiths and Steyvers 2004)

$$P(z_i = j|\mathbf{z}_{\neg i}, w_i, \boldsymbol{\alpha}, \boldsymbol{\beta}) \propto \frac{C^{KW}_{w_{\neg i},j} + \beta_{w_{di},j}}{\sum_{v=1}^{W}(C^{KW}_{v,j} + \beta_{v,j})} \times \frac{C^{KD}_{d_{\neg i},j} + \alpha_{d,j}}{\sum_{k=1}^{K}(C^{KD}_{d,k} + \alpha_{d,k})}, \quad (10.2)$$

where $C^{KW}_{w_{\neg i},j}$ is the number of times word w is assigned to topic j, not including the current token instance i; and $C^{KD}_{d_{\neg i},j}$ is the number of times topic j is assigned to some word token in document d, not including the current instance i. From this distribution, i.e. $p(z_i|\mathbf{z}_{\neg i}, \mathbf{w})$, a topic is sampled and stored as the new topic assignment for this word token. After a sufficient number of sampling iterations, the approximated posterior can be used to get estimates of ϕ and θ by examining the counts of word assignments to topics and topic occurrences in documents.

Given the direct estimate of topic assignments z for every word, it is important to obtain its relation to the required parameters Θ and Φ. This is achieved by sampling new observations based on the current state of the Markov chain (Steyvers and Griffiths 2005). Thus, estimates $\hat{\Theta}$ and $\hat{\Phi}$ of the word – topic and topic – document distributions can be obtained from the count matrices

$$\hat{\phi}_{ik} = \frac{C^{WK}_{i,k} + \beta_{i,k}}{\sum_{v=1}^{W}(C^{WK}_{v,k} + \beta_{v,k})}, \qquad \hat{\theta}_{dk} = \frac{C^{DK}_{d,k} + \alpha_{d,k}}{\sum_{j=1}^{K}(C^{DK}_{d,j} + \alpha_{d,j})}. \quad (10.3)$$

Gibbs sampling has been empirically tested to determine the required length of the burn-in phase, the way to collect samples, and the stability of inferred topics (Griffiths and Steyvers 2004; Heinrich 2005; Steyvers and Griffiths 2005).

10.3.3   Online latent Dirichlet allocation (OLDA)

OLDA is an online version of the LDA model that is able to process text streams (AlSumait et al. 2008). The OLDA model considers the temporal ordering information and assumes that the documents arrive in discrete time slices. At each time slice t of a predetermined size ε, e.g. an hour, a day, or a year, a stream of documents, $S^t = \{d_1, \ldots, d_{D^t}\}$, of variable size, D^t, is received and ready to be processed. A document d received at time t is represented as a vector of word tokens, $\mathbf{w}_d^t = \{w_{d1}^t, \ldots, w_{dN_d}^t\}$. Then, an LDA topic model with K components is used to model the newly arrived documents. The generated model, at a given time, is used as a prior for LDA at the successive time slice, when a new data stream is available for processing (see Figure 10.2 for an illustration). The hyperparameters β can be interpreted as the prior observation counts on the number of times words are sampled from a topic before any word from the corpus is observed (Steyvers and Griffiths 2005), bishop. So, the count of words in topics, resulting from running LDA on documents received at time t, can be used as the priors for the $t + 1$ stream.

Thus, the per-topic distribution over words at time t, $\Phi_k^{(t)}$, is drawn from a Dirichlet distribution governed by the inferred topic structure at time $t - 1$ as follows:

$$\Phi_k^{(t)} | \boldsymbol{\beta}_k^{(t)} \sim Dirichlet(\boldsymbol{\beta}_k^{(t)})$$
$$\sim Dirichlet(\omega \hat{\Phi}_k^{(t-1)}), \qquad (10.4)$$

where $\hat{\Phi}_k^{(t-1)}$ is the frequency distribution of a topic k over words at time $t - 1$ and $0 < \omega \leq 1$ is an *evolution tuning parameter* that is introduced to control the evolution rate of the model. Since the Dirichlet hyperparameters determine the smoothness degree of the priors, it is important to control its effect and to balance

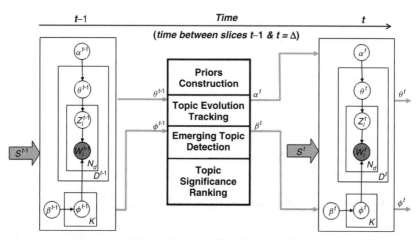

Figure 10.2   A flowchart of OLDA.

between the weight of the past and current semantics in the inference process according to the homogeneity and the evolution rate of the domain's thematic structure. This sequential model is expanded in Section 10.5 to allow data-driven semantic embedding from a wider range of previous models.

Given the definition of $\boldsymbol{\beta}^{(t)}$ in expression (10.4), the topic distributions in consecutive models are aligned so that the evolution of topics in a sequential corpus is captured. For example, if a topic distribution at time t corresponds to a particular theme, then the distribution that has the same ID number in the consecutive models will relate to the same theme, assuming that it appears consistently over time. Thus, the inferred word distribution of topic k at time t can be considered a drifted description of the latent variable k at time $t-1$. The drift is driven by the natural evolution of the topic which includes the changes that occur in the terminology and/or in the interactions with other topics. To model this evolution, an *evolutionary matrix*, $\mathbf{B}_k^{(t)}$, is constructed to capture the evolution of each topic k at each time epoch t within a *sliding history window*, δ. This is given as follows:

$$
\mathbf{B}_k = \begin{pmatrix} \phi_1^{t-\delta} & \cdots & \phi_1^{(t-1)} & \phi_1^{(t)} \\ \phi_2^{t-\delta} & \cdots & \phi_2^{(t-1)} & \phi_2^{(t)} \\ \vdots & \vdots & \vdots & \vdots \\ \phi_{W^{(t)}}^{t-\delta} & \cdots & \phi_{W^{(t)}}^{(t-1)} & \phi_{W^{(t)}}^{(t)} \end{pmatrix},
$$
(10.5)

where each entry $B_k(v, t)$ is the weight of word v under topic k at time t.[1] Thus, working with the evolutionary matrix will allow for tracking the drifts of existing topics, detection of emerging topics, and visualizing the data in general.

Thus, the generative model for time slice t of the proposed OLDA model can be summarized as follows:

1. For each topic $k = 1, \ldots, K$:

   (a) Compute $\boldsymbol{\beta}_k^{(t)} = \omega \hat{\Phi}_k^{(t-1)}$.

   (b) Generate a topic $\Phi_k^{(t)} \sim Dirichlet(\cdot | \boldsymbol{\beta}_k^{(t)})$.

2. For each document, $d = 1, \ldots, D^{(t)}$:

   (a) Draw $\Theta_d^{(t)} \sim Dirichlet(\cdot | \boldsymbol{\alpha}^{(t)})$.

   (b) For each word token, w_{di}, in document d:

        i. Draw $z_i^{(t)}$ from multinomial $\Theta_d^{(t)}$; $(p(z_i^{(t)} | \boldsymbol{\alpha}_d^{(t)}))$.

        ii. Draw $w_{di}^{(t)}$ from multinomial $\Phi_{z_i}^{(t)}$; $p(w_{di}^{(t)} | z_i^{(t)}, \boldsymbol{\beta}_{z_i}^{(t)})$.

---

[1] New observed terms at time t are assumed to have 0 count in ϕ for all topics in previous streams.

Maintaining the models' priors as Dirichlet is essential to simplify the inference problem by making use of the conjugacy property of Dirichlet and multinomial distributions. In fact, by tracking the history as prior patterns, the data likelihood and, hence, the posterior inference of LDA are left the same. Thus, implementing Gibbs sampling in Equation (10.2) in OLDA is straightforward. The main difference of the online approach is that the sampling is performed over the current stream only. This makes the time complexity and memory usage of OLDA efficient and practical. In addition, the β under OLDA are constructed from historic observations rather than fixed values.

10.3.4   Illustrative example

The LDA and OLDA models can be illustrated by generating artificial data from a known topic model and applying the topic models to check whether the data is able to infer the original generative structure. To illustrate the LDA model, six sets of documents are generated from three topic distributions that are equally weighted. Table 10.1 shows the dictionary and topic distributions of the data.

For each set, 16 documents of size 16 word tokens, on average, are generated. After the word assignment vector, $\mathbf{z}$, is randomly initialized, LDA is trained over the documents with the number of components K equal to the true number of components, i.e. K is set to 3. Table 10.2 gives the word – topic correlation counts of LDA averaged over the six sets of documents after 50 iterations of Gibbs sampling. It can be seen that the LDA model is able to correctly estimate the density of each topic.

Table 10.1   Topic distributions of simulated data. Each column is a multinomial distribution of a topic over the dictionary.

Topic	k_1 33%	k_2 34%	k_3 33%
Dictionary↓	$p(w_i|k_1)$	$p(w_i|k_2)$	$p(w_i|k_3)$
river	0.37	0	0
stream	0.41	0	0
bank	0.22	0.28	0
money	0	0.3	0.07
loan	0	0.2	0
debt	0	0.12	0
factory	0	0	0.33
product	0	0	0.25
labor	0	0	0.25
news	0.05	0.05	0.05
reporter	0.05	0.05	0.05

Table 10.2   The frequency distributions of topics discovered by LDA from the static simulated data with K equal to 3.

Topic	T_1 29.8%	T_2 35.5%	T_3 34.7
Dictionary	$f(w_i|T_1)$	$f(w_i|T_2)$	$f(w_i|T_3)$
river	0	0	78
stream	0	0	93
bank	0	56	71
money	0	103	0
loan	0	56	0
debt	0	28	0
factory	85	0	0
production	73	0	0
labor	61	0	0
news	3	19	15
reporter	10	15	14

Table 10.3   Topic distributions of dynamic simulated data over three streams. The rule (—) indicates that the corresponding word or topic has not yet emerged.

Stream	$t = 1$			$t = 2$			$t = 3$		
Topic	k_1 40%	k_2 60%	k_3 0%	k_1 40%	k_2 50%	k_3 10%	k_1 30%	k_2 40%	k_3 30%
Dictionary↓	$p(w_i|k_j)$			$p(w_i|k_j)$			$p(w_i|k_j)$		
river	0.2	0	—	0.4	0	0	0.37	0	0
stream	0.4	0	—	0.2	0	0	0.41	0	0
bank	0.3	0.35	—	0.25	0.36	0.1	0.22	0.28	0
money	0	0.3	—	0	0.24	0	0	0.3	0.07
loan	0	0.25	—	0.05	0.22	0.1	0	0.2	0
debt	—	—	—	0	0.08	0	0	0.12	0
factory	—	—	—	0	0	0.37	0	0	0.33
product	—	—	—	0	0	0.33	0	0	0.25
labor	—	—	—	—	—	—	0	0	0.25
news	0.05	0.05	—	0.05	0.05	0.05	0.05	0.05	0.05
reporter	0.05	0.05	—	0.05	0.05	0.05	0.05	0.05	0.05

Given the same dictionary, three streams of documents are generated from evolving descriptions of topics to demonstrate the OLDA model. Table 10.3 shows the distributions of topics in the three time epochs. Topic 3 emerges as a new topic at the second time epoch. In addition to the new terms introduced by

Table 10.4   Topics discovered by OLDA from dynamic simulated data.

$t = 1$		$t = 2$		$t = 3$	
ID	Topic distribution	ID	Topic distribution	ID	Topic distribution
1	news reporter	1	news reporter	1	reporter news
2	bank	2	bank	2	bank
3	money loan	3	money loan debt	3	money loan debt
4	stream river	4	river stream	4	river stream
5	bank news	5	bank factory production	5	production factory labor

topic 3, a number of terms such as debt and labor gradually emerge. The weight (importance) of topics also varies between the streams. The OLDA topic model is trained on the corresponding documents of each stream with K set to 5. At each time epoch, OLDA is trained on the currently generated documents only. Table 10.4 lists the highest important words under each topic of the evolving simulated data that were discovered by OLDA with K set to 5 at each time epoch. After 50 iterations of Gibbs sampling on each stream, OLDA converged to aligned topic models that correspond to the true topic densities and evolution.

Another observation stems from the setting of K, i.e. the number of components. When K is set to the true number of topics, the topic distributions included some common words in addition to the semantically descriptive ones, see for example the words *news* and *reporter* in topics T_1, T_2, and T_3 in Table 10.2. When K is increased to 5, the topics became more focused as the common words are mapped into individual topics, see topics 1 and 2 in Table 10.3.

10.4   Embedding external semantics from Wikipedia

This section investigates the role of embedding semantics from a source by enhancing the generative process of the model parameters. Such human-defined concept databases provide a natural source of semantics that can provide useful knowledge regarding the hidden thematic structure of the data. We model external knowledge using Wikipedia (Wikipedia 2009). Wikipedia is currently considered the richest online encyclopedia, which consists of a huge number of categorized and consistently structured documents. After the identification of related Wikipedia concepts, LDA is applied to learn a model of the topics discussed in the corresponding Wikipedia articles. The learned topics represent priors about the available knowledge that will be embedded in the inference process of the LDA model to enhance the discovered topics from the text data, which will be referred to hereafter as the test documents.

10.4.1   Related Wikipedia articles

In this work, each Wikipedia article is represented by its title and considered as a single concept. Since Wikipedia includes a large variety of concepts and domains, it is important to use the most related articles to the test documents in order to ensure semantic relatedness and, hence, enhance the inferred model. The related Wikipedia articles are defined to be all Wikipedia concepts that are mentioned in a preset number of test documents, ρ. This is done by searching for the title of the Wikipedia article in the test documents. The threshold value ρ controls the number of Wikipedia articles, $\mathcal{D}$, to be retrieved and, hence, the amount of noise that is allowed to be included in the generative model.

10.4.2   Wikipedia-influenced topic model

After the identification of related Wikipedia concepts, LDA is applied to learn the topics that are discussed in the corresponding Wikipedia articles. In particular, LDA learns two Wikipedia distributions, the topic – word distribution ϕ and the topic – document distribution θ, from

$$\phi_{ik} = \frac{C_{w_i,k}^{WK} + \beta_i}{\sum_{v=1}^{W} C_{v,k}^{WK} + \beta_v}, \qquad \theta_{mk} = \frac{C_{m,k}^{DK} + \alpha_k}{\sum_{j=1}^{K} C_{m,j}^{DK} + \alpha_j}, \qquad (10.6)$$

where m is the index of the Wikipedia article. Within the related Wikipedia articles, $C_{i,k}^{WK}$ is the number of times word i is assigned to topic k and $C_{m,k}^{DK}$ is the number of times topic k is assigned to some word token in Wikipedia article m.

The prior distributions ϕ and θ are then updated into posteriors using the test documents. Specifically, the topic – word distribution ϕ is updated to a new $\hat{\phi}$, and a new topic – document distribution $\hat{\theta}$ is learned from scratch using the test documents

$$\hat{\phi}_{ik} = \frac{C_{w_i,k}^{WK} + \underline{C}_{w_i,k}^{WK} + \beta_i}{\sum_{v=1}^{V} C_{v,k}^{WK} + \underline{C}_{v,k}^{WK} + \beta_v}, \qquad \hat{\theta}_{dk} = \frac{\underline{C}_{d,k}^{DK} + \alpha_k}{\sum_{j=1}^{K} \underline{C}_{d,j}^{DK} + \alpha_j}, \qquad (10.7)$$

where d is the index of the test document, $\underline{C}_{v,k}^{WK}$ is the number of times word v is assigned to topic k, and $\underline{C}_{d,k}^{DK}$ is the number of times topic k is assigned to some word in test document d. Hence, the generative process of the test documents is influenced by the Wikipedia topic model.

10.5   Data-driven semantic embedding

When a topic is observed at a certain time, it is more likely to appear in the future with a similar distribution over words. Unlike general data mining techniques, such an assumption is trivial in the area of TM. It is widely acceptable, for instance, to consider the documents and the words in the documents to be statis-tically dependent. Once a word occurs in a document, it is likely to occur again. Consequently, a similar implication can be made about the topic distribution over time. Despite their natural drifts, the underlying themes of any domain are, in general, consistent. Hence, incorporating prior knowledge about the underlying semantics would eventually enhance the identification and description of topics in the future. In this section, the role of previously discovered topics in inferring future semantics in text streams is investigated under the framework of OLDA topic modeling. A detailed version of the proposed approach can be found in AlSumait et al. (2009).

OLDA is extended to enable semantic embedding in three major directions. First, instead of generating the topic parameters based on the most recently estimated model, the history window is set to incorporate more models in the parameter generation process. Second, the contribution of the semantic history in the inference process is controlled by assigning different weights to different time epochs. Lastly, given the evolutionary matrices of topics defined in Equation (10.5), the priors can be generated using a weighted linear combination of the semantics extracted from all the models that fall within the history window. These three factors are further explained in the following subsections.

10.5.1   Generative process with data-driven semantic embedding

To incorporate inferred semantics from past data, the proposed approach considers all the topic – word distributions learned within a sliding history window, δ, when constructing the current priors. As a result, OLDA can provide alternatives for full, short, or intermediate memory of history.

Given the sliding history window of size c, $1 < c \leq t$, the weight of past models in the prior construction can be controlled by defining a vector of evo-lution tuning parameters ω, instead of the single parameter in expression (10.4). The evolution tuning vector can be used to control the weights of individual models as well as the total weight of history with respect to new semantics. The setting depends mainly on the homogeneity of the data and on the evolution rate of the domain.

The overall influence of history in topic estimation is an important factor that can effect the semantic description of the data. For example, some text repositories, like the scientific literature, persistently introduce novel ideas and, as a consequence, topic distributions change faster compared to other datasets. On the other hand, a great part of the news in news feeds, like sports, stock markets, and weather, are steady over time. Thus, for such consistent topic structures, assigning a higher weight for historic information, compared to the weight of current observations, would improve topic prediction, while the settings should be reversed in fast evolving datasets.

By adjusting the total weight of history, i.e. $\sum_{c=1}^{\delta} \omega_c$, the OLDA model provides a direct way to deploy and tune the influence of history in the inference process. If the total history weight is equal to one, this would (relatively) balance the weights of historic and current observations. When the total weight of history is less (greater) than one, the historic semantic has less (more) influence than the semantic of the current stream.

Thus, given the sliding window δ, the history weight vector $\boldsymbol{\omega}$, and the evolutionary matrix of topic $k\mathbf{B}_k^{(t)}$, as defined in Equation (10.5), the parameters of topic k at time t can be determined by a weighted mixture of the topic's past distributions

$$\boldsymbol{\beta}_k^{(t)} = \mathbf{B}_k^{(t-1)}\boldsymbol{\omega} \tag{10.8}$$

$$= \hat{\Phi}_k^{(t-\delta)}\omega_1 + \cdots + \hat{\Phi}_k^{(t-2)}\omega_{\delta-1} + \hat{\Phi}_k^{(t-1)}\omega_\delta. \tag{10.9}$$

Given the equality in Equation (10.8), the per-topic distribution over words at time t, $\Phi_k^{(t)}$, is drawn from a Dirichlet distribution governed by the evolutionary matrix of the topic as follows:

$$\Phi_k^{(t)}|\boldsymbol{\beta}_k^{(t)} \sim Dirichlet(\boldsymbol{\beta}_k^{(t)})$$
$$\sim Dirichlet(\mathbf{B}_k^{(t-1)}\boldsymbol{\omega}). \tag{10.10}$$

By updating the priors as described above, the structure of the model is kept simple, as all the historic knowledge patterns are printed in the priors rather than in the structure of the graphical model itself. In addition, the learning process on the new stream of data starts from what has been learned so far, rather than starting from arbitrary settings that do not relate to the underlying distributions.

10.5.2   OLDA algorithm with data-driven semantic embedding

An overview of the proposed OLDA algorithm with semantic embedding is shown in Algorithm 8. In addition to the text streams, $S^{(t)}$, the algorithm takes as input the sliding history window size δ, weight vector $\boldsymbol{\omega}$, and fixed Dirichlet values, a and b, for initializing the priors α and β, respectively, at time slice 1. Note that b is also used to set the priors of new words that appear for the first time in any time slice. The output of the algorithm is the generative models and the evolution matrices $\mathbf{B}_k$ for all topics.

Algorithm 8 – OLDA with semantic embedding

1: INPUT: b; a; δ; $\boldsymbol{\omega}$; Δ; $S^{(t)}$, $t = \{1, 2, 3 \dots\}$
2: $t = 1$
3: **loop**
4:     New text stream $S^{(t)}$ is received after time delay equal to Δ
5:     **if** $t = 1$ **then**
6:         $\boldsymbol{\beta}_k^{(t)} = b, k \in \{1, \dots, K\}$
7:     **else**

8:      $\beta_k^t = \mathbf{B}_k^{t-1}\omega$, $k \in \{1, \ldots, K\}$
9:   **end if**
10:    $\alpha_d^{(t)} = a$, $d = 1, \ldots, D^{(t)}$
11:    initialize $\Phi^{(t)}$ and $\theta^{(t)}$ to zeros
12:    initialize topic assignment, $\mathbf{z}^{(t)}$, randomly for all word tokens in $S^{(t)}$
13:    $[\Phi^{(t)}, \Theta^{(t)}, \mathbf{z}^{(t)}] = \text{GibbsSampling}(S^{(t)}, \beta^{(t)}, \alpha^{(t)})$
14:    **if** $t < \delta$ **then**
15:        $\mathbf{B}_k^t = \mathbf{B}_k^{(t-1)} \cup \hat{\Phi}_k^{(t)}$, $k \in \{1, \ldots, K\}$
16:    **else**
17:        $\mathbf{B}_k^t = \mathbf{B}_k^{(t-1)}(1 : W^{(t)}, 2 : \delta) \cup \hat{\Phi}_k^{(t)}$, $k \in \{1, \ldots, K\}$
18:    **end if**
19: **end loop**

___

10.5.3 Experimental design

LDA with semantic embedding is evaluated in the problem domain of document modeling. *Perplexity* is a canonical measure of goodness that is used in language modeling. It evaluates the generalization performance of the model on previously unseen documents. Lower perplexity means a better generalization performance and, hence, a better estimation of density. Formally, for a test set of M documents, the perplexity is (Blei et al. 2003)

$$perplexity(D_{test}) = \exp\left\{ -\frac{\sum_{d=1}^{M} \log p(\mathbf{w}_d)}{\sum_{d=1}^{M} N_d} \right\}. \qquad (10.11)$$

We tested OLDA under different configurations of historic semantic embedding. A summary of the conducted models and their parameter settings are listed in Table 10.5. The window size, δ, was set to values from 0 to 5. The OLDA model with history window of size 0 ignores the history and processes the text stream using a fixed symmetric Dirichlet prior. Under such a model, the estimation is influenced by the semantics of the current stream only. This model, named OLDAFixed, and the OLDA model with $\delta = 1$ are considered as baselines to which the rest of the tested models are compared. To compute the perplexity at every time instance, the documents of the next stream are used as the test set of the model currently generated.

All models were run for 500 iterations and the last sample of the Gibbs sampler was used for evaluation. The number of topics, K, is fixed across all the streams. K, a, and b are set to 50, $50/K$, and 0.01, respectively. All experiments are run on a 2 GHz Pentium M-processor laptop using the MATLAB Topic Modeling Toolbox, authored by Mark Steyvers and Tom Griffiths.[2] The two datasets used in our experiments for the OLDA model with historic semantic embedding are described below.

___

[2] The Topic Modeling Toolbox is available at: http://psiexp.ss.uci.edu/research/programs data/toolbox.htm

Table 10.5   Name and parameter settings of OLDA models. The * indicates that the model was applied on the data.

Reuters	NIPS	Model name	δ	ω
*	*	OLDAFixed	0	NA($\beta = 0.05$)
*	*	1/ω(1)	1	1
*	*	2/ω(1)	2	1, 1
*		2/ω(0.8)	2	0.2, 0.8
*	*	2/ω(0.7)	2	0.3, 0.7
*	*	2/ω(0.6)	2	0.4, 0.6
*	*	2/ω(0.5)	2	0.5, 0.5
*	*	3/ω(1)	3	1, 1, 1
*	*	3/ω(0.8)	3	0.05, 0.15, 0.8
*	*	3/ω(0.7)	3	0.1, 0.2, 0.7
*		3/ω(0.6)	3	0.15, 0.25, 0.6
*	*	3/ω(0.33)	3	0.33, 0.33, 0.34
*	*	4/ω(1)	4	1, 1, 1, 1
	*	4/ω(0.9)	4	0.01, 0.03, 0.06, 0.9
*		4/ω(0.8)	4	0.03, 0.07, 0.1, 0.8
*	*	4/ω(0.7)	4	0.05, 0.1, 0.15, 0.7
*		4/ω(0.6)	4	0.05, 0.15, 0.2, 0.6
*	*	4/ω(0.25)	4	0.25, 0.25, 0.25, 0.25
	*	5/ω(1)	5	1, 1, 1, 1, 1
	*	5/ω(0.7)	5	0.05, 0.05, 0.1, 0.15, 0.7
	*	5/ω(0.6)	5	0.05, 0.1, 0.15, 0.2, 0.6
*	*	5/ω(0.2)	5	0.2, 0.2, 0.2, 0.2, 0.2

Reuters-21578.[3] The corpus consists of newswire articles classified by topic and ordered by their date of issue. There are 90 categories with some articles classified in multiple topics. For our experiments, only articles with at least one topic were kept for processing. For data preprocessing, stop words were removed while the remaining words were down-cased and stemmed to their root source. The resulting dataset consists of 10 337 documents, 12 112 unique words, and a total of 793 936 word tokens. For simplicity, we partitioned the data into 30 slices and considered each slice as a stream.

NIPS dataset.[4] The NIPS set consists of the full text of 13 years of the proceedings from 1988 to 2000 of the Neural Information Processing Systems (NIPS) Conference. The data was preprocessed for down-casing, removing stop words and numbers, and removing those words appearing less than five times in the corpus. The dataset contains 1740 research papers, 13 649 unique words, and 2 301 375 word tokens in total. The set is divided into 13 streams based on the year of publication.

---

[3] The original dataset is available to download from the UCI Knowledge Discovery in Databases Archive: http://archive.ics.uci.edu/ml/.

[4] The original dataset is available at the NIPS Online Repository: http://nips.djvuzone.org/txt.html.

10.5.4   Experimental results

Wikipedia-influenced LDA was run on nine subsets of the Reuters dataset which correspond to the first nine streams. The perplexity of a model was computed using the successive stream as the test set. Figure 10.3 shows the perplexity of Wikipedia-influenced LDA compared to the corresponding models that were trained on the Reuters documents only. It can be seen that the perplexity of LDA with Wikipedia articles is lower in five out of the nine models. We believe that the higher perplexity in some cases with Wikipedia is due to the unstructured approach used to partition the data, which does not guarantee the representation of all the classes in each stream. Thus, any document in the test set that belongs to a new class would eventually increase the perplexity. However, when this factor is neutralized, incorporating external knowledge from Wikipedia does improve the performance.

To test the data-driven semantic embedding, OLDA was first run on the Reuters dataset. It was found that by increasing the window size, δ, OLDA resulted in lower perplexity than the baselines. Figure 10.4 plots the perplexity of OLDA and OLDAFixed at every stream of Reuters under different settings of window size, δ, and the weight vector, ω, was fixed on $1/\delta$. The figure clearly shows that embedding semantics enhanced the document modeling performance. In addition, incorporating semantics from more models, i.e. using a window size greater than 1, further improves the perplexity with respect to OLDA with short memory ($\delta = 1$).

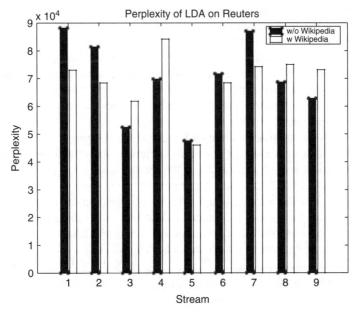

Figure 10.3   Perplexity of OLDA on Reuters with and without Wikipedia articles.

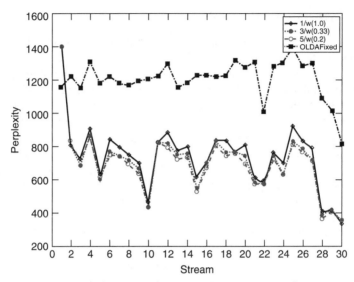

Figure 10.4   Perplexity of OLDA on Reuters for various window sizes compared to OLDAFixed.

Testing with NIPS resulted in a slightly different behavior. When ω was fixed, increasing the window size did show a reduction in the model's perplexity, compared to OLDA with short memory. This is illustrated in Figure 10.5. The larger the window, the lower the perplexity of the model. Nonetheless, the OLDA model only showed improvements with respect to OLDAFixed when the window size was larger than 3. In addition to the window size, previous experiments on NIPS suggested the effect of the total weight of history in estimating the topical semantics of heterogeneous and fast evolving domains like scientific research (AlSumait et al. 2008). The experiments explained next provide evidence of such a justification. Nonetheless, it is worth mentioning here that the OLDA model outperforms OLDAFixed in its ability to automatically detect and track the underlying topics.

To investigate the role of the total history weight, we tested OLDA on NIPS and Reuters under a variety of ω settings. Figure 10.6 shows the average perplexity of OLDA with δ fixed at 2 and the total sum of ω set to 0.05, 0.1, 0.15, 0.2, and 1 for both datasets. Both baselines, OLDAFixed and OLDA with short memory, are also shown. We found that the contribution of history in NIPS is completely opposite to that in Reuters. While increasing the weight for history resulted in a better topical description of Reuters news, lower perplexities were reported with NIPS only for topic models that assign a lower weight for history. In fact, the history weight and perplexity in NIPS (Reuters) are negatively (positively) correlated.

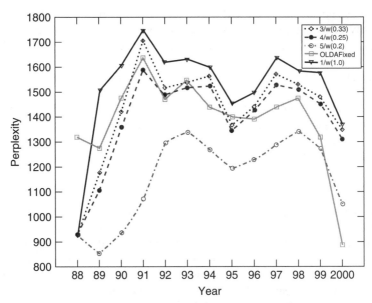

Figure 10.5   Perplexity of OLDA on NIPS for various window sizes compared to OLDAFixed.

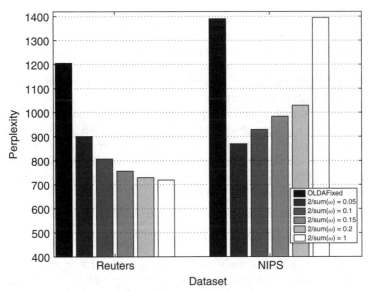

Figure 10.6   Average perplexity of OLDA on Reuters and NIPS under different weights of history contribution compared to OLDA with fixed β.

Reuters' documents span a short period of time while the streams of NIPS are yearly based. As a result, the Reuters' topics are homogeneous and more stable. So, letting the current generative model be heavily influenced by the past topical structure will eventually result in a better description of the data. On the other hand, although there is a set of predefined publication domains in NIPS, like algorithms, applications, and visual processing, these topics are very broad and interrelated. Furthermore, research papers usually cover more topics and continuously introduce novel ideas and topics. Hence, the influence of previous semantics should not exceed the topical structure of the present.

10.6   Related work

The problem of embedding semantic information within the document representation and/or distance metrics has recently been investigated intensively in the domain of text classification and clustering (e.g. AlSumait and Domeniconi (2008), Cristianini et al. (2002)). However, the problem of embedding semantic information within the generative model and the inference process of LDA topic modeling is a new research area. Very recently (Andrzejewski et al. 2009), domain knowledge has been implemented in the form of must-link and cannot-link primitives about the word compositions that should have high or low probability in the topics. These primitives are incorporated in LDA using a mixture of Dirichlet tree priors.

A number of papers in the literature have used LDA topic modeling to represent some kind of semantic embedding. In the domain of text segmentation, the work in Sun et al. (2008) used an LDA-based Fisher kernel to measure text semantic similarity between blocks of documents in the form of latent semantic topics that were previously inferred using LDA. The kernel is controlled by the number of shared semantics and word co-occurrences. Phrase discovery is another area that aims at identifying phrases (n-grams) in text. Wang et al. (2007) presented a topical n-gram model that automatically identified feasible n-grams based on the context that surround it. Moreover, there are some research efforts to incorporate prior knowledge from large universal datasets, like Wikipedia. Phan et al. (2008) built a classifier on both a small set of labeled documents and an LDA topic model estimated from Wikipedia.

10.7   Conclusion and future work

In this chapter, the effect of embedding semantic information in the framework of probabilistic topic modeling is investigated. In particular, static and online LDA topic models are first introduced and two directions to embed semantics within their inference process are defined. The first direction updates the topical structure based on prior knowledge that is learned from Wikipedia. The second approach constructs the parameters based on the topical semantics that have been inferred by the past generated models.

This work can be extended in many directions. LDA with external semantic embedding can be used to build an unsupervised classifier that can effectively group documents based on their content with no need for labeled training documents. In addition, it can be extended to work online on text streams and using an evolving external knowledge. The effect of the embedded historic semantics on detecting emerging and/or periodic topics constitutes future work.

References

AlSumait L and Domeniconi C 2008 Text clustering with local semantic kernels. In *Survey of Text Mining: Clustering, Classification, and Retrieval* (ed. Berry M and Castellanos M) 2nd edn Springer.

AlSumait L, Barbará D and Domeniconi C 2008 Online LDA: Adaptive topic model for mining text streams with application on topic detection and tracking. *Proceedings of the IEEE International Conference on Data Mining*.

AlSumait L, Barbará D and Domeniconi C 2009 The role of semantic history on online generative topic modeling. *Proceedings of the Workshop on Text Mining, held in conjunction with the SIAM International Conference on Data Mining*.

Andrzejewski D, Zhu X and Craven M 2009 Incorporating domain knowledge into topic modeling via Dirichlet forest priors *Proceedings of the International Conference on Machine Learning*.

Blei D, Ng A and Jordan M 2003 Latent Dirichlet allocation. *Journal of Machine Learning Research* **3**, 993–1022.

Cristianini N, Shawe-Taylor J and Lodhi H 2002 Latent semantic kernels. *Journal of Intelligent Information Systems* **18**(2–3), 127–152.

Deerwester S, Dumais S, Furnas G, Landauer T and Harshman R 1990 Indexing by latent semantic analysis. *Journal of the American Society for Information Science* **41**(6), 391–407.

Griffiths T and Steyvers M 2004 Finding scientific topics. *Proceedings of the National Academy of Sciences*, pp. 5228–5235.

Heinrich G 2005 *Parameter Estimation for Text Analysis*. Springer.

Hofmann T 1999 Probabilistic latent semantic indexing. *Proceedings of the 15th Conference on Uncertainty in Artificial Intelligence*.

Mimno D and McCallum A 2007 Organizing the OCA: Learning faceted subjects from a library of digital books. *Proceedings of the Joint Conference on Digital Libraries*.

Minka T and Lafferty J 2002 Expectation-propagation for the generative aspect model. *Proceedings of the 18th Conference on Uncertainty in Artificial Intelligence*.

Papadimitriou C, Tamaki H, Raghavan P and Vempala S 2000 Latent semantic indexing: A probabilistic analysis. *Journal of Computer and System Sciences* **61**(2), 217–235.

Phan X, Nguyen L and Horiguchi S 2008 Learning to classify short and sparse text and web with hidden topics from large-scale data collections. *International WWW Conference Committee*.

Salton G 1983 *Introduction to Modern Information Retrieval*. McGraw-Hill.

Steyvers M and Griffiths T 2005 Probabilistic topic models. In *Latent Semantic Analysis: A Road to Meaning* (ed. Landauer T, McNamara D, Dennis S and Kintsch W) Lawrence Erlbaum Associates.

Story R 1996 An explanation of the effectiveness of latent semantic indexing by means of a Bayesian regression model. *Information Processing and Management* **32**(3), 329–344.

Sun Q, Li R, Luo D and Wu X 2008 Text segmentation with LDA-based Fisher kernels. *Proceedings of the Association for Computational Linguistics*.

Wang X, McCallum A and Wei X 2007 Topical n-grams: Phrase and topic discovery, with an application to information retrieval. *Proceedings of the 7th IEEE International Conference on Data Mining*.

Wei X and Croft B 2006 LDA-based document models for ad-hoc retrieval. *Proceedings of the Conference on Research and Development in Information Retrieval*.

White C 2005 Consolidating, accessing and analyzing unstructured data.

Wikipedia 2009 Wikipedia: The free encyclopedia.

Index

adaptive threshold setting, 132

centroid, 82, 83
chat rooms, 151
chi-square statistic, 173
clustering
   constrained, 81
   pairwise constrained, 81
confusion matrix, 99
constrained optimization, 83
constraints
   cannot-link, 82, 84, 87, 93,
       102
   instance-level, 81
   must-link, 82, 87, 94, 102
   must–must-link, 99
correlation function, 178
cybercrime, 161
   cyberbullying, 150
   cyberpredators, 150

dataset
   CISI collection, 99
   Cranfield collection, 99
   Medlars collection, 99
   MPQA corpus, 15
   NIPS, 198
   Reuters, 198
distance
   Bregman, 89
   Kullback–Leibler, 82, 89
   reversed Bregman, 90
   squared Euclidean, 82
divergence

Bregman, 89
Kullback–Leibler, 82
reverse Bregman, 81

event types, 167
external prior-knowledge, 193

FeatureLens, 113
first variation, 83
function
   closed, proper, convex, 89
   distance-like, 83
FutureLens, 113

gain ratio, 66
Gaussian algorithm, 173
Gaussian distribution, 133

history flow, 110
hypothesis test, 172

information gain, 66
information retrieval (IR), 22, 95
isolating language, 23, 32

k-means, 81
   batch, 83
   constrained, 101
   incremental, 82, 83, 86, 100
   quadratic, 82, 95
   quadratic batch, 82
   spherical, 81
   spherical batch, 95, 96
   spherical constrained, 101
   spherical incremental, 97, 99

k-means, (*Continued*)
 spherical with constraints, 82
keyword, 3, 170
 applications, 3, 4, 15
 extraction methods, 4, 5
 keyphrase, 3, 5
 metrics, 17, 18
 variants, 16

latent Dirichlet allocation, 186
 generative process, 187
 inference, 187
  Gibbs sampling, 188
 online LDA, 189
latent morpho-semantic analysis
   (LMSA), 32
 with term alignments
   (LMSATA), 33
latent semantic analysis, 185
 probabilistic, 185
latent semantic analysis (LSA), 22
 with term alignments (LSATA),
   30
latent semantic indexing, 72
LDA, *see* latent Dirichlet
   allocation, 186
log-entropy term weighting, 26
LSI, *see* latent semantic indexing,
   72
luring communication, 154

mean
 arithmetic, 90
misbehavior detection, 152
multi-parallel corpus, 21
multilingual document clustering,
   21
multilingual LSA, 25
multilingual precision at five
   documents (MP5), 25

NMF, *see* nonnegative matrix
   factorization, 60
NMF-BCC, 74
NMF-LSI, 72

nonnegative matrix factorization,
   60
 alternating least squares
   algorithm, 62
 classification, 70
 initialization, 65
 multiplicative update algorithm,
   62
novelty mining, 130
NP-hard problem, 83

online
 communities, 150
 luring, 154
 victimization, 149

pairwise mutual information (PMI),
   30
PARAFAC2, 28
partition, 83
 nextFV (II), 84
 quality, 83
PDDP, 99, 100
penalty, 82, 84, 87, 88, 93, 96, 99,
   102
power method, 30
precision at one document (P1), 24
predatory behavior, 159

RAKE, 5
 algorithm, 6, 7
 evaluation, 9, 10, 15
 input parameters, 6
RSS, 169

SEASR, 111
semantic embedding
 data driven, 194
 external, 193
sentiment tracking, 111
sexual
 exploitation, 149
 predation, 150
singular value decomposition, 22
Sinkhorn balancing, 30
smoka, 81, 92

constrained, 101
SMT, *see* statistical machine
    translation, 30
social networking, 149
statistical machine translation, 30
stoplist, 6, 10
  generation, 11
  stopwords, 5
SVD, *see* singular value
    decomposition, 22
swine flu, 178
synthetic language, 23, 32

tag
  cloud, 108
  crowd, 108
temporal profiles, 171
tensors, 27
text stream, 169

TextArc, 111
topicality, 180
transitive closure, 88, 94, 98
TREC novelty track data, 138
Tucker1, 27

UIMA, 111

vector space model, 22, 184
vocabulary
  indexing, 4
  term frequency, 12
  term selection, 4
VSM, *see* vector space model,
    22

Wikipedia, *see* external prior
    knowledge, 193
Wordle, 108